CONSTRUCTION SCHEDULING,

COST OPTIMIZATION, AND MANAGEMENT

A New Model Based on Neurocomputing and Object Technologies

Hojjat Adeli

Asim Karim

The Ohio State University

CRC Press
Taylor & Francis Group
Boca Raton London New York

CRC Press is an imprint of the
Taylor & Francis Group, an **informa** business

A SPON PRESS BOOK

CRC Press
Taylor & Francis Group
6000 Broken Sound Parkway NW, Suite 300
Boca Raton, FL 33487-2742

First issued in paperback 2019

ISBN-13: 978-0-415-24417-6 (hbk)
ISBN-13: 978-0-367-86380-7 (pbk)

Publisher's Note
This book has been prepared from camera-ready copy provided by the author.

British Library Cataloguing in Publication Data
A catalogue record for this book is available from the British Library

Library of Congress Cataloging in Publication Data
Adeli, Hojjat.
Construction scheduling, cost optimization, and management
H. Adeli and A. Karim.
1.Construction industry—Management. 2. Production scheduling.
3. Building—Cost control. I. Karim, A. (Asim) II. Title.
TH438.4 .A33 2001
634'.068--dc21
00-062904

**Visit the Taylor & Francis Web site at
http://www.taylorandfrancis.com**

**and the CRC Press Web site at
http://www.crcpress.com**

DEDICATED TO

Nahid, Anahita, Amir Kevin, Mona, and Cyrus Dean
Adeli

and

Salim and Abida
Karim

CONTENTS

PREFACE

The primary purpose of this book is to present an entirely new approach to management and scheduling of construction projects overcoming the limitations of existing methods. We start from ground zero with a most general mathematical formulation for scheduling and management of construction projects with the goal of minimizing the direct construction cost. The construction direct cost optimization problem is then solved by the robust neural dynamics model of Adeli and Park. An object-oriented information model is presented based on the new construction scheduling model, laying the foundation for a new generation of flexible, powerful, maintainable, and reusable software system for the construction scheduling problem.

In order to demonstrate the practicality of the new computational and information models for management and scheduling of actual construction projects, they have been implemented in a new generation software system, called CONSCOM (for CONstruction Scheduling, Cost Optimization, and Change Order Management).

Some of the unique features of CONSCOM non-existent in CPM-based models are described through examples. It must be pointed out that CONSCOM is not just a software system but represents a new technology for management and scheduling of construction projects based on advanced computational, neurocomputing, and object technologies.

The current prevailing design and construction practice is to complete the design before the construction is started. In other words, design and construction are treated as two independent and separate activities. Integration of design and construction through the emerging field of *concurrent* or *collaborative engineering* provides a number of advantages, and an opportunity to advance the two fields of construction engineering and structural engineering significantly. CONSCOM with its change order management capability is particularly suitable for use in a concurrent engineering environment. Successful application of concurrent engineering in the construction industry should be based on effective integration of the construction management and scheduling with the design process. The other essential prerequisite for such an integration is a tool to automate the complex process of engineering design. A chapter in the book is devoted to this subject. Finally, for the sake of completeness, fundamentals of project planning, scheduling, and management, and the ubiquitous industry standard Critical Path Method (CPM) are also presented in the book.

ACKNOWLEDGMENT

The work presented in this book was partially sponsored by the *Ohio Department of Transportation* and *Federal Highway Administration.* Parts of the work presented in this book were published by authors in several journal articles in *Computer-Aided Civil and Infrastructure Engineering* and *Engineering, Construction, and Architectural Management* (published by Blackwell Publishers), *Journal of Construction Engineering and Management* and *Journal of Structural Engineering* (published by American Society of Civil Engineers), and *Thin-Walled Structures* (published by Elsevier), as noted in the list of references. Chapter 10 of the book is based on the article: Adeli, H. and Wu, M., "Regularization Neural Network for Construction Cost Estimation," *Journal of Construction Engineering and Management*, ASCE, Vol. 124, No. 1, 1998, and reprinted by the permission of the publisher.

ABOUT THE AUTHORS

Hojjat Adeli is currently Professor of Civil and Environmental Engineering and Geodetic Science, Director of Knowledge Engineering Lab, and a member of the Center for Cognitive Science at The Ohio State University. A contributor to 53 different scholarly journals, he has authored over 340 research and scientific publications in diverse areas of engineering, computer science, and applied mathematics. He has authored/co-authored nine pioneering books. His recent books are *Machine Learning – Neural Networks, Genetic Algorithms, and Fuzzy Systems*, John Wiley and Sons, 1995, *Neurocomputing for Design Automation*, CRC Press, 1998, *Distributed Computer-Aided Engineering*, CRC Press, 1999, *High-Performance Computing in Structural Engineering*, CRC Press, 1999, and *Control, Optimization, and Smart Structures – High-Performance Bridges and Buildings of the Future*, John Wiley and Sons, 1999. He has also edited 12 books including *Intelligent Information Systems*, IEEE Computer Society, 1997. He is the Editor-in-Chief of two research journals, **Computer-Aided Civil**

and Infrastructure Engineering which he founded in 1986 and **Integrated Computer-Aided Engineering** which he founded in 1993. He has been a Keynote/Plenary Lecturer at 38 international computing conferences held in 28 different countries. On September 29, 1998, he was awarded a patent for a *"Method and Apparatus for Efficient Design Automation and Optimization, and Structures Produced Thereby"* (United States Patent Number 5,815,394) (with a former Ph.D. student). He is the recipient of numerous academic, research, and leadership awards, and honors, and recognition. In 1998, he was awarded the **University Distinguished Scholar Award** by The Ohio State University *"in recognition of extraordinary accomplishment in research and scholarship"*, and the Senate of the General Assembly of State of Ohio passed a resolution honoring him as an *"Outstanding Ohioan."* He is listed in 26 different Who's Who's and archival biographical listings such as *Two Thousand Notable Americans, The Directory of Distinguished Americans, Five Hundred Leaders of Influence,* and *Two Thousand Outstanding People of the 20th Century.* He has been an organizer or a member of organization/scientific/program committee of over 160 conferences held in 44 different countries. His research has been sponsored by 20 different organizations including government funding agencies such as the *National Science Foundation, U.S. Air Force Flight Dynamics Laboratory,* and *U.S. Army Construction Engineering Research Laboratory, Federal Highway Administration,* state funding agencies such as the *Ohio Department of*

INTRODUCTION

Concurrent engineering technology has been developed substantially in automotive and other manufacturing applications. It is now finding applications in the construction industry in both the U.S. (El-Bibany and Paulson, 1999; Pena-Mora and Hussein, 1999) and Japan (Kaneta *et al.*, 1999). The current prevailing practice is to complete the design before the construction is started. But, changes in the design might be necessary to improve the product or project even after the construction has already begun. Successful application of concurrent engineering in the construction industry should be based on effective integration of the construction management and scheduling with the design process (Adeli, 1999, 2000). Two essential prerequisites for such an integration are a tool to automate the complex process of engineering design and a tool for construction scheduling, cost optimization and change order management.

Automation of design of large one-of-a-kind civil engineering systems is a challenging problem due partly to the open-ended nature of the problem and partly to the highly nonlinear constraints that can baffle optimization algorithms (Adeli, 1994). Optimization of large and complex engineering systems is particularly challenging in terms of convergence, stability, and efficiency. Recently, Adeli and Park (1995a, 1998) developed a robust neural dynamics optimization model for automating the complex process of engineering design and applied it to very large-scale problems including the minimum weight design of a 144-story superhighrise building structure with more than 20,000 members.

The construction project schedule is an important document in the construction industry. The construction project participants, such as the contractor and the owner, use the construction schedule to plan, monitor, and control project work. The goal is the completion of the project within budget and time, and to the satisfaction of all project participants. In recent years the construction schedule has increasingly been used as a legal document in resolving disputes and verifying claims among the project participants. Further, change order claims are communicated, verified, and studied through construction schedules. Therefore, the value of the construction schedule cannot be overstated. For its effectiveness, every construction schedule must have two essential characteristics. First, it must be based on an accurate model of the construction project. Second, it must provide features necessary for project control and management. A software system for generating and maintaining such schedules is therefore highly desirable.

A construction schedule is traditionally defined as the timetable of the execution of tasks in a project. Resources are assigned to the tasks before the project is scheduled. Thus, resources are handled separately and independently of the time. This results in a cost model that is disconnected from the time model thus making cost and time control difficult and imprecise. Further, the scheduled times of the tasks in the project are not updateable in a structured manner to reflect changes that have occurred since the project started. This makes project monitoring, control, and change order management cumbersome and difficult.

In recent years, the use of construction scheduling software has made the use of schedules in the construction industry widespread. However, the underlying modeling technology used and the process of control and management has not progressed significantly over the years. The critical path method (CPM), developed in the late 1950s, is still used despite its documented schedule-modeling shortcomings, particularly for projects involving repetitive tasks (Adeli and Karim, 1997). The limitations and shortcomings of the existing software systems used in practice are also recognized by the construction industry. For example, a majority of the members of the Associated General Contractors of America are dissatisfied with the critical path method (CPM) (Mattila and Abraham, 1998). CPM is an easy technique but is based on many simplifying assumptions in modeling construction projects. Despite these concerns and shortcomings, the CPM is still widely used and none of the other methods presented

over the years has gained widespread acceptance. Newer techniques
have failed because they are not a significant improvement over the
CPM. Further, newer techniques must be made available to the
construction industry in a software system that is easy to use.

To address these shortcomings was our motivation to develop an
advanced general construction scheduling model and implement it in
a prototype software system called CONSCOM. By utilizing recent
advances in computational science and engineering and information
technology an advanced and powerful scheduling model is presented
for accurate and realistic project planning and management.

An overview of neural network applications in civil engineering
is presented in Chapter 2. The neural dynamics optimization model
of Adeli and Park is described in Chapter 3 in the context of
engineering design optimization and using the design of steel cold-
formed beams as an example. Chapter 4 provides an introduction to
project planning, scheduling, and management. The CPM is also
described in this section. A general mathematical formulation is
presented in Chapter 5 for scheduling of construction projects.
Various scheduling constraints are expressed mathematically. The
construction scheduling is posed as an optimization problem where
the project direct cost is minimized for a given project duration. The
neural dynamics model of Adeli and Park is adapted for solution of
this problem in Chapter 6.

Chapter 7 presents an object-oriented information model for
construction scheduling, cost optimization, and change order

management based on the new mathematical construction scheduling model. The model provides support for schedule generation and review, cost estimation, and cost-time trade-off analysis, and is implemented as an application development *framework*. In Chapters 8 and 9, a specific implementation of the model is described in a prototype software system called CONSCOM (CONstruction, Scheduling, Cost Optimization, and Change Order Management) using the Microsoft Foundation Class (MFC) library. Key software design concepts and ideas necessary for the development of extensible, compatible, and maintainable software systems and used in CONSCOM are presented in Chapter 8. Unique features of CONSCOM and its integrated software management environment are described in Chapter 9. This chapter also includes an example illustrating the use of CONSCOM for construction cost minimization and change order management. Finally, a regularization neural network model for construction cost estimation is presented in Chapter 10.

Overview of Neural Networks in Civil Engineering

2.1 INTRODUCTION

Artificial neural networks (ANN) are a functional abstraction of the biological neural structures of the central nervous system (Aleksander and Morton, 1993; Rudomin *et al.*, 1993; Arbib, 1995; Anderson, 1995; Bishop, 1995). They are powerful pattern recognizers and classifiers. They operate as black box, model-free, and adaptive tools to capture and learn significant structures in data. Their computing abilities have been proven in the fields of prediction and estimation, pattern recognition, and optimization (Adeli and Hung, 1995; Golden, 1996; Mehrotra *et al.*, 1997; Adeli and Park, 1998; Haykin, 1999). They are particularly suitable for problems too complex to be modeled and solved by classical mathematics and traditional procedures. A neural network can be trained to recognize a particular pattern or learn to perform a particular task. The approach is particularly attractive for *hard-to-learn* problems and when there is no formal underlying theory for the solution of the problem. Engineering design and image recognition are two such problems (Adeli and Hung, 1995).

The first journal article on civil/structural engineering applications of neural network was published by Adeli and Yeh (1989). In this chapter we present an overview of neural network articles published in archival research journals since then. The great majority of civil engineering applications of neural networks are based on the simple backpropagation (BP) algorithm. Applications of other recent more powerful and efficient neural networks models are also reviewed. Recent works on integration of neural networks with other computing paradigms such as genetic algorithm, fuzzy logic, and wavelet to enhance the performance of neural network models are presented.

One of the reasons for popularity of the neural network is the development of the simple error backpropagation (BP) training algorithm (Rumelhart *et al.*, 1986) which is based on a gradient-descent optimization technique. The BP algorithm is now described in many textbooks (Adeli and Hung, 1995; Mehrotra *et al.*, 1997; Topping and Bahreininejad, 1997; Haykin, 1999) and the unfamiliar reader can refer to any one of them. A review of the BP algorithm with suggestions on how to develop practical neural network applications is presented by Hegazy *et al.* (1994). The great majority of the civil engineering application of neural networks is based on the use of the BP algorithm primarily due to its simplicity. Training of a neural network with a supervised learning algorithm such as BP means finding the weights of the links connecting the nodes using a set of training examples. An error function in the form of the sum of the squares of the errors between the actual outputs from the training set and the computed outputs is minimized iteratively. The learning or training rule specifies how the weights are modified in each iteration.

2.2   CONSTRUCTION ENGINEERING

2.2.1   Construction Scheduling and Management

Adeli and Karim (1997b) present a general mathematical formulation for scheduling of construction projects and apply it to the problem of highway construction scheduling. Repetitive and non-repetitive tasks, work continuity considerations, multiple-crew strategies, and the effects of varying job conditions on the performance of a crew can be modeled. An optimization formulation is presented for the construction project scheduling problem with the goal of minimizing the direct construction cost. The nonlinear optimization is then solved by the neural dynamics model of Adeli and Park (1996a). This model is described in detail in Chapters 5 and 6.

Karim and Adeli (1999a) present an object-oriented information model for construction scheduling, cost optimization, and change order management based on the new neural network-based construction scheduling model of Adeli and Karim (1997b). The model has been implemented in a prototype software system called CONSCOM (CONstruction Scheduling, Cost Optimization, and

Change Order Management) using Microsoft Foundation Class library under the Windows environment (Karim and Adeli, 1999b). The development of the object model and the CONSCOM software system are described in Chapters 7 to 9.

2.2.2 Construction Cost Estimation

Williams (1994) attempts to use the BP algorithm for predicting changes in construction cost indexes for one month and six months ahead but concludes that *"the movement of the cost indexes is a complex problem that cannot be predicted accurately by a BP neural network model."* Automating the process of construction cost estimation based on objective data is highly desirable not only for improving the efficiency but also for removing the subjective questionable human factors as much as possible. The costs of construction materials, equipments, and labor depend on numerous factors with no explicit mathematical model or rule for price prediction.

Adeli and Wu (1998) point out that *"highway construction costs are very noisy and the noise is the result of many unpredictable factors such as human judgment factors, random market fluctuations, and weather conditions."* They also discuss the problem of over-fitting data, noting that *"because of the noise in the data, a perfect fit usually is not the best fit,"* and under-fitting resulting in poor generalization. Adeli and Wu (1998) present a regularization neural network model and architecture for estimating the cost of construction projects. The new computational model is based on a solid mathematical foundation making the cost estimation consistently more reliable and predictable. The new cost estimation model is presented in Chapter 10.

2.2.3   Resource Allocation and Scheduling

Mohammad *et al.* (1995) formulate the problem of optimally allocating available yearly budget to bridge rehabilitation and replacements projects among a number of alternatives as an optimization problem using the Hopfield network (Hopfield, 1982, 1984). Savin *et al.* (1996, 1998) also discuss the use of a discrete-time Hopfield net in conjunction with an augmented Lagrangian multiplier optimization algorithm for construction resource leveling. Elazouni *et al.* (1997) use the BP algorithm to estimate the construction resource requirements at the conceptual design stage and apply the model for the construction of concrete silo walls.

Senouci and Adeli (2001) present a mathematical model for resource scheduling considering project scheduling characteristics generally ignored in prior research, including precedence relationships, multiple crew-strategies, and time-cost trade-off. Previous resource scheduling formulations have traditionally focused on project duration minimization. The new model considers the total project cost minimization. Furthermore, resource leveling and resource-constrained scheduling are performed simultaneously. The model is solved using the neural dynamics optimization model of Adeli and Park (1996a).

2.2.4   Construction Litigation

Disputes and disagreements between the contractor and the owner for reasons such as misinterpretation of the contract, changes made by the owner or the contractor, differing site and weather conditions, labor problems, and unexpected delays can lead to litigation. Arditi *et al.* (1998) use neural networks to predict the outcome of construction litigation. They use the outcomes of circuit and appellate court decisions to train the network and report a successful prediction rate of 67 percent for the *"extremely complex data structure of court proceedings."* A comparison of the neural network approach with Case-Based Reasoning (CBR) for the same problem is presented by Arditi and Tokdemir (1999).

2.2.5   Other Applications of BP and Other Neural Network Models in Construction Engineering and Management

Moselhi *et al.* (1991) were among the first to realize the potential applications of neural networks in construction engineering. They present an application of the BP algorithm for optimum markup estimation under different bid conditions. They use a small set of 10 bid situations to train the system but report up to 30,000 iterations for the BP algorithm to converge with a small error. The BP algorithm has also been used for selection of vertical concrete formwork supporting walls and columns for a building site (Kamarthi *et al.*, 1992), for estimating construction productivity (Chao and Skibniewski, 1994; Sonmez and Rowings, 1998), for markup estimation using knowledge acquired from contractors in Canada and the U.S. (Hegazy and Moselhi, 1994), for evaluation of new construction technology acceptability (Chao and Skibniewski, 1995), for selection of horizontal concrete formwork to support slabs and roofs (Hanna and Senouci, 1995), and for measuring the level of organization effectiveness in a construction firm (Sinha and McKim, 2000).

Murtaza and Fisher (1994) describe the use of neural networks for decision making about construction modularization. Yeh (1995) uses a combination of simulated annealing (Kirkpatrick *et al.*, 1983) and Hopfield neural network (Hopfield, 1982, 1984) to solve the construction site layout problem. Kartam (1996) uses neural networks to determine optimal equipment combinations for earthmoving operations. Pompe and Feelders (1997) use neural networks to predict corporate bankruptcy. Li *et al.* (1999) discuss rule extractions from a neural network trained by the BP algorithm for construction mark-up estimation in order to explain how a particular recommendation is made.

2.3   STRUCTURAL ENGINEERING

2.3.1   Pattern Recognition and Machine Learning in Structural Analysis and Design

Adeli and Yeh (1989) present a model of machine learning in engineering design based on the concept of internal control parameters and perceptron (Rosenblatt, 1962). A perceptron is defined as a four-tuple entity (sensors to receive inputs, weights to be multiplied by the sensors, a function collecting all of the weighted data to produce a proper measurement on the impact of the observed phenomenon, and a constant threshold) and the structural design problem is formulated as a perceptron without hidden units. They apply the model to design of steel beams.

Vanluchene and Sun (1990) demonstrate potential applications of the BP algorithm (Rumelhart *et al.*, 1986) in structural engineering by presenting its application to three problems: a simple beam load location problem involving pattern recognition, the cross-section selection of reinforced concrete beams involving typical design decisions, and analysis of a simply supported plate showing how numerically complex solutions can be estimated quickly with the neural networks approach.

Hajela and Berke (1991) demonstrate neural networks can be used for rapid re-analysis for structural optimization. Hung and Adeli (1991a) present a model of machine learning in engineering design, called PERHID, based on the concept of perceptron learning algorithm (Rosenblatt, 1962, Adeli and Yeh, 1989) with a two-layer neural network. PERHID has been constructed by combining perceptron with a single layer AND neural net. Extending this research, Hung and Adeli (1994a) present a neural network machine learning development environment using the object-oriented programming paradigm (Yu and Adeli, 1991, 1993).

(1994) for design of simple trusses, by Hoit *et al.* (1994) for equation renumbering in finite element analysis of structures to improve profile and wavefront characteristics, by Rogers (1994) for fast approximate structural analysis in a structural optimization program, by Mukherjee and Deshpande (1995a&b) for the preliminary design of structures, by Abdalla and Stavroulakis (1995) to predict the behavior of semi-rigid connections in steel structures from experimental moment rotation curves for single-angle and single-plate beam-column connections, by Turkkan and Srivastava (1995) to predict the steady state wind pressure profile for air-supported cylindrical and hemispherical membrane structures, by Mukherjee *et al.* (1996) to predict the buckling load of axially loaded columns based on experimental data, by Papadrakakis *et al.* (1996) for structural reliability analysis in connection with the Monte Carlo simulation, by Anderson *et al.* (1997) to predict the bi-linear moment-rotation characteristics of the minor-axis beam-to-column connections based on experimental results, by Szewczyk and Noor (1996, 1997) for sensitivity and nonlinear analysis of structures, by Kushida *et al.* (1997) to develop a concrete bridge rating system, by Hegazy *et al.* (1998) to model the load-deflection behavior, concrete strain distribution at failure, reinforcing steel strain distribution at failure, and crack-pattern formation of concrete slabs, by Chuang *et al.* (1998) to predict the ultimate load capacity of pin-ended reinforced concrete columns, by Stavroulakis and Antes (1998) for crack identification in steady state elastodynamics, by Cao *et al.*

(1998) to identify loads on aircraft wings modeled approximately as a cantilever beam subjected to a set of concentrated loads, by Mathew *et al.* (1999) for analysis of masonry panels under biaxial bending, and by Jenkins (1999) for structural re-analysis of two-dimensional trusses.

Biedermann (1997) investigates the use of the BP neural networks to represent heuristic design knowledge such as how to classify the members of a multistory frame into a limited number of groups for practical purposes (design fabrication groups). Cattan and Mohammadi (1997) use the BP algorithm to relate the subjective rating of bridges based on visual inspection of experienced bridge inspectors to the analytical rating based on detailed structural analyses under standard live loads as well as the bridge parameters. They conclude that *"neural networks can be trained and used successfully in estimating a rating based on bridge parameters."*

Adeli and Park (1995c) present application of counterpropagation neural networks (CPN) with competition and interpolation layers (Hecht-Nielsen, 1987a, b, 1988) in structural engineering. A problem with the CPN algorithm is the arbitrary trial-and-error selection of the learning coefficients encountered in the algorithm. The authors propose a simple formula for the learning coefficients as a function of the iteration number and report excellent convergence results. The CPN algorithm is used to predict elastic critical lateral torsional buckling moment of wide-flange steel beams (W shapes) and the moment-gradient coefficient for doubly and singly symmetric steel beams subjected to end moments. The latter is a complex stability analysis problem requiring a large neural network with 4224 links, extensive numerical analysis, and management of a large amount of data. It took less than 30 iterations to train the large CPN network in both competition and interpolation layers using 528 training instances. Compared with the BP algorithm, they found superior convergence property and a substantial decrease in the processing time for the CPN algorithm with the proposed formula for the learning coefficients.

The computation of an effective length factor (K) is complicated but essential for design of members in compression in steel frame structures. The present AISC codes for design of steel structures (AISC, 1995, 1998) present simplified alignment charts for determining the effective length factor. Duan and Chen (1989) and Kishi *et al.* (1997) have pointed out the gross underestimation (leading to an unsafe design) and overestimation (leading to an overly conservative design) of the alignment charts for different boundary conditions. Hung and Jan (1999a) describe a variation of the cerebellar model articulation controller (CMAC), used mostly in the control domain, for predicting the effective length factor (K) for columns in unbraced frames. They conclude that the results obtained from the neural network model are more accurate than those obtained from the AISC alignment charts.

In the finite element analysis of structures the relationship between the loads and displacements is represented by the structure or global stiffness matrix. A neural network can be trained to perform the same task. Solution of the simultaneous linear equations including the stiffness matrix is the most time-consuming part of any large-scale finite element analysis. To speed up this step of the finite element analysis, neural networks can be used to create domain specific equation solvers utilizing the knowledge of a particular domain such as highway bridges. But, neural networks can provide only an approximate solution where an "*exact*" solution is usually required. Consolazio (2000) proposes combining neural networks with iterative equation-solving techniques such as a preconditioned conjugate gradient (PCG) algorithm (Adeli and Kumar, 1999). In particular, he uses the BP neural network algorithm to compute approximate displacements at each iteration, while the overall PCG algorithm steers convergence to the exact solution. The neural network part of the algorithm improves the efficiency of the algorithm by (a) providing a good initial solution and (b) playing the role of the preconditioner in the PCG algorithm. The author applies the method to finite element analysis of flat-slab highway bridges and concludes the neural network to be an effective method for accelerating the convergence of iterative methods. Use of neural network in finite element analysis is also discussed by Li (2000).

2.3.2 Design Automation and Optimization

Automation of design of large one-of-a-kind civil engineering systems is a challenging problem due partly to the open-ended nature of the problem and partly to the highly nonlinear constraints that can baffle optimization algorithms (Adeli, 1994). Optimization of large and complex engineering systems is particularly challenging in terms of convergence, stability, and efficiency. Most of the neural networks research has been done in the area of pattern recognition and machine learning (Adeli and Hung, 1995). Neural network computing can also be used for optimization (Berke *et al.*, 1993).

Adeli and Park (1995a) present a neural dynamics model for optimal design of structures by integrating penalty function method, Lyapunov stability theorem, Kuhn-Tucker conditions, and the neural dynamics concept. A pseudo-objective function in the form of a Lyapunov energy functional is defined using the exterior penalty function method. The Lyapunov stability theorem guarantees that solutions of the corresponding dynamic system (trajectories) for arbitrarily given starting points approach an equilibrium point without increasing the value of the objective function. The robustness of the model was first verified by application to a linear structural optimization problem, the minimum weight plastic design of lowrise planar steel frames (Park and Adeli, 1995). Optimization algorithms are known to deteriorate with the increase in size and complexity of the problem. The significance of the new optimization model is that it provides the optimum design of large structures with thousands of members subjected to complicated and discontinuous constraints with excellent convergence results.

In order to achieve automated optimum design of realistic structures subjected to actual constraints of commonly-used design codes such as the American Institute of Steel Construction (AISC) Allowable Stress Design (ASD) and Load and Resistance Factor Design (LRFD) specifications (AISC, 1995, 1998), Adeli and Park (1995b, 1996a) developed a hybrid CPN-neural dynamics for discrete optimization of structures consisting of commercially available sections such as the wide-flange (W) shapes used in steel structures. The computational models are shown to be highly stable and robust and particularly suitable for design automation and optimization of large structures no matter how large the size of the problem is, how irregular the structure is, or how complicated the constraints are. For their innovative landmark work the authors were awarded a patent by the *U.S. Patent and Trademark Office* on September 29, 1998 (United States Patent Number 5,815,394).

An important advantage of cold-formed steel is the greater flexibility of cross-sectional shapes and sizes available to the structural steel designer. The lack of standard optimized shapes, however, makes the selection of the most economical shape very difficult if not impossible. This task is further complicated by the complex and highly nonlinear nature of the rules that govern their design. Adeli and Karim (1997a) present a general mathematical formulation and computational model for optimization of cold-formed steel beams. The nonlinear optimization problem is solved by adapting the robust neural dynamics model of Adeli and Park (1996a). The basis of design can be the AISI ASD or LRFD specifications (AISI, 1996, 1997). The computational model is applied to three different commonly-used types of cross-sectional shapes: hat, I, and Z shapes. The computational model was used to perform extensive parametric studies to obtain the global optimum design curves for cold-formed hat, I, and Z shape steel beams based on the AISI code to be used directly by practicing design engineers (Karim and Adeli, 1999d, e, 2000). Results from this work are presented in Chapter 3.

Optimization of space structures made of cold-formed steel is complicated because an effective reduced area must be calculated for members in compression to take into account the non-uniform distribution of stresses in thin cold-formed members due to torsional/flexural buckling. The effective area varies not only with the level of the applied compressive stress but also with its width-to-thickness ratio. Tashakori and Adeli (2001) present optimum (minimum weight) design of space trusses made of cold-formed steel shapes in accordance with the AISI specifications (AISI, 1996, 1997) using the neural dynamics model of Adeli and Park (1996a). The model has been used to find the minimum weight design for several space trusses commonly used as roof structures in long-span commercial buildings and canopies, including a large structure with 1548 members with excellent convergence results.

Arslan and Hajela (1997) discuss counterpropagation neural networks in decomposition-based optimal design. Parvin and Serpen (1999) discuss a procedure to solve an optimization problem with a single-layer, relaxation type recurrent neural network, but do not present solution to any significant structural design problem.

2.3.3   Structural System Identification

Masri *et al.* (1993) describe neural networks as a powerful tool for identification of structural dynamic systems. Chen *et al.* (1995b) use the BP algorithm for identification of structural dynamic models. The authors indicate *"great promise in structural dynamic model identification by using neural network"* based on simulation results for a real multi-story building subjected to earthquake ground motions. The BP algorithm is also used by Yun and Bahng (2000) for substructural identification and estimating the stiffness parameters of two-dimensional trusses and frames. Huang and Loh (2001) propose a neural network-based model for modeling and identification of a discrete-time nonlinear hysteretic system during strong earthquakes. They use two-dimensional models of a three-story frame and a real bridge in Taiwan subjected to several earthquake accelerograms to validate the feasibility and reliability of the method for estimating the changes in structural response under different earthquake events.

2.3.4   Structural Condition Assessment and Monitoring

Wu *et al.* (1992) discuss the use of the BP algorithm for detection of structural damage in a three-story frame with rigid floors. The damage is defined as a reduction in the member stiffness. Elkordy *et al.* (1994) question the reliability of the traditional methods for structural damage diagnosis and monitoring which relies primarily on visual inspection and simple on-site tests. They propose a structural damage monitoring system for identifying the damage associated with changes in structural signatures using the BP algorithm. For training they used experimental results from a shaking table as well as numerical results from a finite element analysis of the structure for strain-mode shapes as the vibrational signatures. They point out that *"analyzing the data obtained from different types of sensors to detect damage is a very complex problem, particularly because of the noise associated with the signals"* and suggest neural networks can diagnose complicated damage patterns and *"can handle noisy and partially incomplete data sets."* Stephens and Vanluchene (1994) describe an approach for assessing the safety condition of structures after the occurrence of a damaging earthquake using multiple quantitative indices and the BP algorithm. They concluded that the neural network model *"generated more reliable assessments than could be obtained using any single indicator or from a linear regression model that utilized all indicators."*

Defining damage as a reduction in the stiffness of structural members, Szewczyk and Hajela (1994) use a CPN network for damage detection in truss and frame structures. They describe the problem as an inverse static analysis problem where the elements of the structure stiffness matrix are found based on experimentally observed response data. Pandey and Barai (1995) describe the use of the BP algorithm for damage detection of steel truss bridge structures. A similar study for vibration signature analysis of steel trusses is discussed in Barai and Pandey (1995). Masri *et al.* (1996) explore the use of neural networks to detect changes in structural parameters during vibrations. Masri *et al.* (2000) describe application of neural networks for a nonparametric structural damage detection methodology based on nonlinear system identification approaches.

High-strength bolts in spliced joints of steel bridges may become loose gradually during their lifetime. This problem has to be detected and corrected during periodic inspection and maintenance of the bridge. Mikami *et al.* (1998) present a system based on the BP algorithm to estimate the residual axial forces of high-strength bolts in steel bridges using the reaction and acceleration waveforms collected by an automatic hammer or looseness detector.

An important issue in structural health monitoring is the selection of the members and locations of the structure to be monitored. Feng and Bahng (1999) use the BP algorithm to estimate the change in the stiffness based on the measured vibration characteristics for damage assessment of reinforced concrete columns retrofitted by advanced composite jackets. Kim *et al.* (2000b) describe a two-stage procedure where in the first stage traditional sensitivity analysis is used to rank and select critical members. In the second stage, the results of the sensitivity analysis and a trained neural network are used to identify the optimal numbers and location of monitoring sensors. The method is applied to two-dimensional trusses and multistory frames.

2.3.5   Structural Control

Active control of structures has been an active area of research in recent years (Adeli and Saleh, 1999). Ghaboussi and Joghataie (1995) present application of neural networks in structural control. A neural network training algorithm, a modified BP algorithm in this case, performs the role of the control algorithm. The structure's response measured at a selected number of points by sensors and the actuator signals are the input to the *neurocontroller*. Its output is the subsequent value of the actuator signal to produce the desired actuator forces. The neurocontroller learns to control the structure after being trained by an emulator neural network. The authors suggest that neurocontrollers are a potentially power tool in structural control problems based on simulation results for a three-story frame with one actuator.

Chen *et al.* (1995a) also describe the use of the BP algorithm in structural control and present simulation results based on the model of an actual multi-story apartment building subjected to recorded earthquake ground motions. The BP algorithm is also used by Tang (1996a) for active control of a single-degree-of-freedom system and by Yen (1996) for vibration control in flexible multibody dynamics. Nikzad *et al.* (1996) compare the performances of a conventional feedforward controller and a neurocontroller based on a modified BP algorithm in compensating the effects of the actuator dynamics and computational phase delay using a two-degree-of-freedom dynamic system and report the latter is *"far more effective"*. Most control algorithms are based on the availability of a complete state vector from measurement. Tang (1996b) uses the BP algorithm as the state vector estimator when only a limited number of sensors are installed in the structure and consequently a complete state vector is not available. Bani-Hani and Ghaboussi (1998) discuss nonlinear structural control using neural networks through numerical simulations on a two-dimensional three-story steel frame considering its inelastic material behavior.

Ankireddi and Yang (1999) investigate the use of neural networks for failure detection and accommodation in structural control problems. They propose a failure detection neural network for monitoring structural responses and detecting performance-reducing sensor failures and a failure accommodation neural network to account for the failed sensors using the Widrow-Hoff training rule (Widrow and Lehr, 1995). Kim *et al.* (2000a) propose an optimal control algorithm using neural networks through minimization of the instantaneous cost function for a single-degree-of-freedom system. Hung *et al.* (2000) describe an active pulse structural control using neural networks with a training algorithm that does not require the trial-and-error selection of the learning ratio needed in the BP algorithm and present simulation results for a small frame.

2.3.6   Finite Element Mesh Generation

In finite element analysis of structures creating the right mesh is a tedious and trial and error process often requiring a high level of human expertise. The accuracy and efficiency of the method rely heavily on the selected mesh. Automatic creation of an effective finite element mesh for a given problem has been an active area of research. Different approaches have been explored in the literature including neural networks. For a given number of nodes and mesh topology, Manevitz *et al.* (1997) use the self-organizing algorithm of Kohonen (1988) to create a near-optimal finite element mesh for a two-dimensional domain using a combination of different types of elements. Bahreininejad *et al.* (1996) explore the application the BP and Hopfield neural networks for finite element mesh partitioning. Pain *et al.* (1999) present a neural network graph partitioning algorithm for partitioning unstructured finite element meshes. First, an automatic graphic coarsening method is used to create a coarse mesh followed by a mean field theorem neural network to perform partitioning optimization.

2.3.7   Structural Material Characterization and Modeling

Ghaboussi *et al.* (1991) describe the use of the BP neural network for modeling behavior of conventional materials such as concrete in the state of plane stress under monotonic biaxial loading. Brown *et al.* (1991) demonstrate the applicability of neural networks to composite material characterization. They use the BP algorithm to predict hygral, thermal, and mechanical properties of composite ply materials. The BP algorithm has also been used for constitutive modeling of concrete (Sankarasubramanian and Rajasekaran, 1996) and viscoplastic materials (Furukawa and Yagawa, 1998).

Ghaboussi *et al.* (1998) present *autoprogressive* training of neural network constitutive models using the global load-deflection response measured in a structural test with application to laminated composites. In their approach a partially trained neural network generates its own training cases through an iterative nonlinear finite element analysis of the test specimen. Yeh (1999) uses the BP algorithm to model the concrete workability in design of high-performance concrete mixture. Neural network is also used to model generalized hardening plasticity (Theocaris and Panagiotopoulos, 1995), alkali-silica reaction of concrete with admixtures (Li *et al.*, 2000), and elastoplasticity (Daoheng *et al.*, 2000).

2.3.8   Parallel Neural Network Algorithms for Large-Scale Problems

The convergence speed of neural network learning models is slow. For large networks several hours or even days of computer time may be required using the conventional serial workstations. A parallel backpropagation learning algorithm has been developed by Hung and Adeli (1993) and implemented on the Cray YMP supercomputer. A parallel processing implementation of the BP algorithm on a Transputer network with application to finite element mesh generation is also presented by Topping *et al.* (1997).

Optimization of large structures with thousands of members subjected to actual constraints of commonly-used design codes requires an inordinate amount of computer processing time and high-performance computing resources (Adeli and Kamal, 1993; Adeli,

1992a&b; Adeli and Soegiarso, 1999). Park and Adeli (1997a) present a data parallel neural dynamics model for discrete optimization of large steel structures implemented on a distributed memory multiprocessor, the massively parallel Connection Machine CM-5 system. The parallel algorithm has been applied to optimization of several highrise and superhighrise building structures including a 144-story steel superhighrise building structure with 20096 members in accordance with the AISC ASD and LRFD codes (AISC, 1995, 1998) and subjected to multiple loading conditions including wind loading according to the Uniform Building Code (UBC, 1997). This is by far the largest structural optimization problem subjected to actual constraints of a widely-used design code ever solved and reported in the literature. Park and Adeli (1997b) present distributed neural dynamics algorithms on the Cray T3D multiprocessor employing the work sharing programming paradigm.

2.4   ENVIRONMENTAL AND WATER RESOURCES ENGINEERING

Karunanithi et al. (1994) demonstrate the use of neural networks for river flow prediction using the cascade-correlation algorithm. The BP algorithm is used by Du et al. (1994) to predict the level of solubilization of six heavy metals from sewage sludge using the bio-leaching process, by Grubert (1995) to predict the flow conditions at the interface of stratified estuaries and fjords, by Kao and Liao (1996) to facilitate the selection of an appropriate facility combination for municipal solid waste incineration, by Tawfik *et al.* (1997) to model stage-discharge relationships at stream gauging locations at the Nile river, by Deo *et al.* (1997) to interpolate the ocean wave heights over short intervals (weekly mean wave heights) from the values obtained by remote sensing techniques and satellites over long durations (a month), and by Liong *et al.* (2000) for water level forecasting in Dhaka, Bangladesh.

Crespo and Mora (1995) describe neural network learning for river streamflow estimation, prediction of carbon dioxide concentration from a gas furnace, and feedwater control system in a boiling water reactor. Basheer and Najjar (1996) use neural networks to model fixed-bed adsorber dynamics. Rodriguez and Serodes (1996) use the BP neural network to estimate the disinfectant dose adjustments required during water re-chlorination in storage tanks based on representative operational and water quality historical data and conclude that the model *"can adequately mimic an operator's know-how in the control of the water quality within distribution systems"*. Maier and Dandy (1997) discuss the use of neural networks for multivariate forecasting problems encountered in the field of water resources engineering including estimation of salinity in a river. Thirumalaiah and Deo (1998) present neural networks for real-time forecasting of stream flows. Flood values during storms are forecast with a lead time of one hour or more using the data from past flood values at a specific location. Deo and Chaudhari (1998) use neural networks to predict tides at a station located in the interior of an estuary or bay.

Gangopadhyay *et al.* (1999) integrate the BP algorithm with a Geographic Information System (GIS) for generation of subsurface profiles and for identification of the distribution of subsurface materials. The model is applied to find the aquifer extent and its parameters for the multi-aquifer system under the city of Bangkok, Thailand. Coulibaly *et al.* (2000) use feedforward and recurrent neural networks for long-term forecasting of potential energy inflows for hydropower operations planning. This is one of the few papers addressing the problem of overfitting in neural network pattern recognition. The authors conclude that *"the neural network-based models provide more accurate forecasts than traditional stochastic models."* Liu and James (2000) use the BP algorithm to estimate the discharge capacity in meandering compound (or two-stage) channels consisting of a main channel flanked by floodplains on one or both sides. Guo (2001) presents a semi-virtual watershed model for small urban watersheds with a drainage area of less than 150 acres using neural networks where the network training and the determination of the matrix of time-dependent weights to rainfall and runoff vectors is guided by the kinematic wave theory.

2.5   TRAFFIC ENGINEERING

Cheu and Ritchie (1995) use three different neural network architectures: multi-layer perceptron, self-organizing feature map, and adaptive resonance theory (ART) model two (ART2), for the identification of incident patterns in traffic data. Faghri and Hua (1995) use ART model one (ART1) to estimate the average annual daily traffic (AADT) including the seasonal factors and compare its performance with clustering and regression methods. They conclude that the neural network model yields better results than the other two approaches. Dia and Rose (1997) use field data to test a multi-layer perceptron neural network as an incident detection classifier. Eskandarian and Thiriez (1998) use neural networks to simulate a driver's function of steering and braking and develop a controller on a moving platform (vehicle) encountering obstacles of various shapes. The system can generalize its learned patterns to avoid obstacles and collision.

The BP neural network is used by Lingras and Adamo (1996) to estimate average and peak hourly traffic volumes, by Ivan and Sethi (1998) for traffic incident detection, by Sayed and Abdelwahab (1998) for classification of road accidents for road improvements, and by Park and Rilett (1999) to predict the freeway link travel times for one through five time periods into the future. Saito and Fan (2000) present an optimal traffic signal timing model that uses the BP algorithm to conduct an analysis of the level of service at a signalized intersection by learning the complicated relationship between the traffic delay and traffic environment at signalized intersections.

2.6   HIGHWAY ENGINEERING

Gagarin *et al.* (1994) discuss the use of radial-Gaussian-based neural network for determining truck attributes such as axle loads, axle spacing, and velocity from strain-response readings taken from the bridges over which the truck is travelling. Eldin and Senouci (1995) describe the use of the BP algorithm for condition rating of roadway pavements. They report very low average error when compared with a human expert determination. Cal (1995) uses the BP algorithm for soil classification based on three primary factors: plastic index, liquid limit water capacity, and clay content. Razaqpur *et al.* (1996) present a combined dynamic programming and Hopfield neural network (Hopfield, 1982, 1984) bridge management model for efficient allocation of a limited budget to bridge projects over a given period of time. The time dimension is modeled by dynamic programming and the bridge network is simulated by the neural network. Roberts and Attoh-Okine (1998) uses a combination of supervised and self-organizing neural networks to predict the performance of pavements as defined by the International Roughness Index.

The BP algorithm is used by Owusu-Ababio (1998) for predicting flexible pavement cracking and by Alsugair and Al-Qudrah (1998) to develop a pavement management decision support system for selecting an appropriate maintenance and repair action for a damaged pavement. Attoh-Okine (2001) uses the self-organizing map or competitive unsupervised learning model of Kohonen (1988) for grouping of pavement condition variables (such as the thickness and age of pavement, average annual daily traffic, alligator cracking, wide cracking, potholing, and rut depth) to develop a model for evaluation of pavement conditions.

2.7   GEOTECHNICAL ENGINEERING

A common method for evaluation of elastic moduli and layer thicknesses of soils and pavements is the seismic spectral-analysis-of-surface-waves (SASW). Williams and Gucunski (1995) use the BP algorithm to perform the inversion of SASW test results. Core penetration test (CPT) measurements are frequently used to find soil strength and stiffness parameters needed in design of foundations. Goh (1995) demonstrates application of the BP algorithm for correlating various experimental parameters and evaluating the CPT calibration chamber test data. The BP algorithm is used by Chikata *et al.* (1998) to develop a system for aesthetic evaluation of concrete retaining walls and by Teh *et al.* (1997) to estimate static capacity of precast reinforced concrete piles from dynamic stress wave data. Juang and Chen (1999) present neural network models for evaluating the liquefaction potential of sandy soils. Use of neural networks to predict the collapse potential of soils is discussed by Juang *et al.* (1999).

After pointing out *"classical constitutive modeling of geomaterials based on the elasticity and plasticity theories suffers from limitations pertaining to formulation complexity, idealization of behavior, and excessive empirical parameters"*, Basheer (2000) proposes neural networks as an alternative for modeling the constitutive hysteresis behavior of soils. He examines several mapping techniques to be used as frameworks for creating neural network models for constitutive response of soils including a hybrid approach that provides high accuracy.

2.8   SHORTCOMINGS OF THE BP ALGORITHM AND OTHER RECENT APPROACHES

2.8.1   Shortcomings of the BP Algorithm

The momentum BP learning algorithm (Rumelhart *et al.*, 1986, Adeli and Hung, 1995) is widely used for training multilayer neural networks for classification problems. This algorithm, however, has a slow rate of learning. The number of iterations for learning an example is often in the order of thousands and sometimes more than one hundred thousands (Carpenter and Barthelemy, 1994). Moreover, the convergence rate is highly dependent on the choice of the values of learning and momentum ratios encountered in this algorithm. The proper values of these two parameters are dependent on the type of the problem (Adeli and Hung, 1994; Yeh, 1998). As such, a number of other neural network learning models have been proposed in recent years. Some of them with applications in civil engineering are reviewed briefly in this section.

2.8.2   Adaptive Conjugate Gradient Neural Network Algorithm

In an attempt to overcome the shortcomings of the BP algorithm, Adeli and Hung (1994) have developed an adaptive conjugate gradient learning algorithm for training of multilayer feedforward neural networks. The Powell's modified conjugate gradient algorithm has been used with an approximate line search for minimizing the system error. The problem of arbitrary trial-and-error selection of the learning and momentum ratios encountered in the momentum backpropagation algorithm is circumvented in the new adaptive algorithm. Instead of constant learning and momentum ratios, the step length in the inexact line search is adapted during the learning process through a mathematical approach. Thus, the new adaptive algorithm provides a more solid mathematical foundation for neural network learning. The algorithm has been applied to the domain of image recognition. It is shown that the adaptive neural networks algorithm has superior convergence property compared with the momentum BP algorithm (Adeli and Hung, 1995).

2.8.3   Radial Basis Function Neural Networks

The radial basis function neural network (RBFNN) learns an input–output mapping by covering the input space with basis functions that transform a vector from the input space to the output space (Moody and Darken, 1989; Poggio and Girosi, 1990). Conceptually, the RBFNN is an abstraction of the observation that biological neurons exhibit a receptive field of activation such that the output is large when the input is closer to the center of the field and small when the input moves away from the center. Structurally, the RBFNN has a simple topology with a hidden layer of nodes having nonlinear basis transfer functions and an output layer of nodes with linear transfer functions (Adeli and Karim, 2000). The most common type of the basis function is Gaussian. Yen (1996) proposes the use of radial basis function networks as a neurocontroller for vibration suppression. Amin *et al.* (1998) use the RBFNN to predict the flow of traffic. Jayawardena and Fernando (1998) present application of the RBFNN for hydrological modeling and runoff simulation in a small catchment and report it is more efficient computationally compared with the BP algorithm.

2.8.4   Other Approaches

Masri *et al.* (1999) propose a stochastic optimization algorithm based on adaptive random search techniques for training neural networks in applied mechanics applications. Castillo *et al.* (2000a) present functional networks where neural functions are learned instead of weights but apply the concept to simple problems such as predicting behavior of a cantilever beam and approximating the differential equation for vibration of a simple single-degree-of-freedom system with spring and viscous damping. Some learning methods in functional networks are presented in Castillo *et al.* (2000b).

2.9 INTEGRATING NEURAL NETWORK WITH OTHER COMPUTING PARADIGMS

2.9.1 Genetic Algorithms

Hung and Adeli (1991b) present a hybrid learning algorithm by integrating genetic algorithm with error backpropagation multilayer neural networks. The algorithm consists of two learning stages. The first learning stage is to accelerate the learning process by using a genetic algorithm with the feedforward step of the BP algorithm. In this stage, the weights of the neural network are encoded on chromosomes as decision variables. The objective function for the genetic algorithm is defined as the average squared system error. After performing several iterations and meeting the stopping criterion, the first learning stage is terminated and the chromosome returning the minimum objective function is considered as the initial weights of the neural network in the second stage. Next, the BP algorithm performs the second learning process until the terminal condition is satisfied.

Moselhi et al. (1993) use the BP neural networks and the genetic algorithm (Adeli and Hung, 1995) to develop a decision-support system to aid the contractors in preparing bids. A parallel genetic/neural network algorithm is also presented by Hung and Adeli (1994b). Jinghui et al. (1996), Hajela and Lee (1997), and Papadrakakis et al. (1998) use the BP algorithm to improve the efficiency of genetic algorithms for structural optimization problems.

Topping *et al.* (1998) present parallel finite element analysis on a MIMD distributed computer. They describe a mesh partitioning technique for planar finite element meshes where a BP neural network is used to find the approximate number of elements within a coarse mesh. The coarse mesh is then divided into several subdomains using a GA optimization approach.

2.9.2   Fuzzy Logic

Adeli and Hung (1993) present a fuzzy neural network learning model by integrating an unsupervised fuzzy neural network classification algorithm with a genetic algorithm and the adaptive conjugate gradient neural network learning algorithm. The learning model consists of three major stages. The first stage is used to classify the given training instances into a small number of clusters using the unsupervised fuzzy neural network classification algorithm. The second stage is a supervised neural network learning model using the classified clusters as training instances. The genetic algorithm is used in this stage to accelerate the whole learning process in the hybrid learning algorithm. The third stage is the process of defuzzification. The hybrid fuzzy neural network learning model has been applied to the domain of image recognition. The performance of the model has been evaluated by applying it to a large scale training example with 2304 training instances.

Hurson *et al.* (1994) discuss use of the fuzzy logic in automating knowledge acquisition in a neural network-based decision support system. Anantha Ramu and Johnson (1995) present a fuzzy logic-BP neural network approach to detect, classify, and estimate the extent of damage from the measured vibration response of composite laminates. Kasperkiewicz *et al.* (1995) use a fuzzy ART neural network (Carpenter *et al.*, 1991) to predict strength properties of high-performance concrete mixes as a factor of six components: cement, silica, superplasticizer, water, fine aggregate, and coarse aggregate.

Furuta *et al.* (1996) describe a fuzzy expert system for damage assessment of reinforced concrete bridge decks using genetic algorithms and neural networks. The goal is to automatically acquire fuzzy production rules through the use of the GA and the BP neural networks. The weights of the links obtained from the neural networks are used in the GA evaluation function to obtain the optimal combination of rules to be used in the knowledge base of the expert system (Adeli, 1988; Adeli and Balasubramanyam, 1988; Adeli, 1990a, b).

Ni *et al.* (1996) present a fuzzy neural network approach for evaluating the stability of natural slopes considering the geological, topographical, meteorological, and environmental conditions that can be described mostly in linguistic terms. Parameters of the neural networks are represented by fuzzy sets (Zadeh, 1970, 1978). Faravelli and Yao (1996) discuss the use of neural networks in fuzzy control of structures. Rajasekaran *et al.* (1996) describe the integration of fuzzy logic and neural networks for prestressed concrete pile diagnosis problem and concrete mix design. Hung and

Jan (1999b) present a fuzzy neural network learning model consisting of both supervised and unsupervised learning and apply it to simply supported concrete and steel beam design problems. Sayed and Razavi (2000) combine fuzzy logic with an adaptive B-spline network to model the behavioral mode choice in the area of transportation planning. They apply the model to a bimodal example for shipment of commodities (rail and Interstate Commerce Commission-regulated motor carriers for shipments over 500 lb).

2.9.3   Wavelets

Neural network models can lose their effectiveness when the patterns are very complicated or noisy. Traffic data collected from loop detectors installed in a freeway system and transmitted to a central station present such patterns. Neural networks have been used to detect incident patterns from non-incident patterns with limited success. The dimensionality of the training input data is high and the embedded incident characteristics are not easily detectable. Adeli and Samant (2000) present a computational model for automatic traffic incident detection using discrete wavelet transform (Samant and Adeli, 2000) and neural networks. Wavelet transform is used for feature extraction, de-noising, and effective preprocessing of data before the adaptive conjugate gradient neural network model of Adeli and Hung (1994) is used to make the traffic incident detection. The authors show that for incidents with duration of more than five minutes, the incident detection model yields a detection rate of nearly 100% and false alarm rate of about 1% for two- or three-lane freeways.

Adeli and Karim (2000) present a new multi-paradigm intelligent system approach to the traffic incident detection problem through integration of fuzzy, wavelet, and neural computing techniques to improve reliability and robustness. A wavelet-based de-noising technique is employed to eliminate undesirable fluctuations in observed data from traffic sensors. Fuzzy c-mean clustering is used to extract significant information from the observed data and to reduce its dimensionality. A radial basis function neural network is developed to classify the de-noised and clustered observed data. The authors report excellent incident detection rates with no false alarms when tested using both real and simulated data.

Liew and Wang (1998) describe application of wavelets for crack identification in structures. Marwala (2000) uses the wavelet transform and neural networks for damage identification in structures.

3

Neural Dynamics Model and its Application to Engineering Design Optimization

---

3.1   INTRODUCTION

Solving engineering problems requires that decisions be made reliably and accurately towards the desired solution. Whether the problem involves design and development, operation and maintenance, or control and management, decisions must be based on a solid model of the system that serves the purpose of the solution. This systematic procedure leads to decisions that are consistently reliable and reproducible. Furthermore, such a procedure can be automated for efficient processing. On the other hand, decisions made somewhat arbitrarily by engineers are open to personal interpretation and inconsistencies.

Most engineering decisions can be posed as an optimization problem where the maximum or minimum of an objective is sought that satisfies all the system constraints. Once a problem is modeled mathematically with objective and constraint functions, mathematical optimization techniques can be used to solve it. Many optimization algorithms are available; however, their applications have been limited mostly to academic problems.

The objectives of this chapter are:

- To introduce the neural dynamics model of Adeli and Park (1995, 1998) as a general model for the solution of complex nonlinear optimization problems.

- To describe how the neural dynamics model can be adapted to a specific optimization problem—in this case, the minimum weight design of simply supported cold-formed steel beams.

- To demonstrate the robustness and practical applicability of the model by presenting parametric studies for the design of hat-shaped cold-formed steel beams.

3.2   COLD-FORMED STEEL DESIGN OPTIMIZATION

An important advantage of cold-formed steel is the greater flexibility of cross-sectional shapes and sizes available to the structural steel designer. Through cold-forming operations, steel sheets, strips or plates can easily be shaped and sized to meet a large variety of design options. Such a large number of design possibilities creates an important challenge: How to choose the most economical cold-formed shape in design of steel structures.

A search of structural engineering and computing journals up to 1996 turned out only one journal article on the optimization of cold-formed steel structures. Seaburg and Salmon (1971) used the direct and gradient search techniques for the minimum weight design of hat-shaped cold-formed light gauge steel members. The design was based on the 1968 American Iron and Steel Institute's (AISI) Specifications for the Design of Cold-Formed Structural Steel Members (AISI, 1968). They presented only one example without any convergence curves. The method used had difficulty finding the optimum depth and did not guarantee that the solution obtained was a local optimum design. Still, considering the highly nonlinear nature of the optimization problem their work should be considered significant, especially noting that it was done some 30 years ago.

3.3 MINIMUM WEIGHT DESIGN OF COLD-FORMED STEEL BEAMS

The optimization problem is defined as:

Minimize

$$W = F(\mathbf{X}) \tag{3.1}$$

subject to

$$g_j^k(\mathbf{X}) \le 0 \quad j = 1, I; k = 1, N_L \tag{3.2}$$

$$\mathbf{X}^L \le \mathbf{X} \le \mathbf{X}^U \tag{3.3}$$

where W is the weight of the structure, $\mathbf{X}$ is the vector of design variables, I is the number of design constraints, N_L is the number of load cases, and $\mathbf{X}^L$ and $\mathbf{X}^U$ are the lower and upper bounds on the

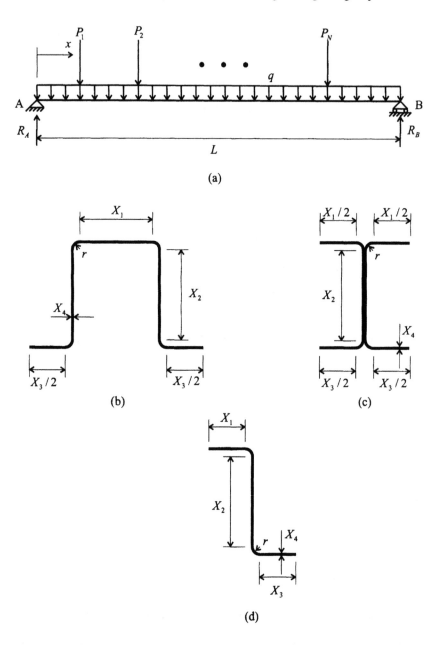

Figure 3.1   Variables involved in the design of hat-, I-, and Z-shaped
cold-formed steel beams

design variables, respectively. In the design of cold-formed steel beams, **X** consists of the cross-sectional dimensions of the members. The exact form of the objective function and the constraints depend on the shape of the section used (Figure 3.1).

The designs are based on two different specifications: (1) The AISI Specification for the Design of Cold-formed Steel Structural Members – 1989 Edition with 1989 Addendum (AISI, 1989), hereafter referred to as the AISI ASD Specification, and (2) the AISI Load and Resistance Factor Specification for the Design of Cold-Formed Steel Structural Members (AISI, 1991), hereafter referred to as the AISI LRFD Specification. The loading on the beams is assumed to be a uniformly distributed load and/or any number of concentrated loads. Full lateral bracing, no lateral bracing, or lateral bracing at any number of specified points along the length of the beam may be assumed.

The beam span and loading is shown in Figure 3.1a. The optimization model presented in this chapter is general and can be applied to any kind of cross section. As examples, however, we consider three commonly-used shapes: hat, I, and Z shapes. The variables involved in their design are shown in Figures 3.1b–d. For these shapes, the weight (objective function) is defined as:

$$W = \rho L X_4 \left[X_1 + X_3 + N_w (X_2 + 0.5\pi r^2) \right] \tag{3.4}$$

where ρ is the unit weight of steel, L is the span length of the beam, N_w is the number of webs (equal to 1 for Z shape and 2 for hat and I shapes), and r, X_1, X_2, X_3, and X_4 are the cross-sectional dimensions identified in Figure 3.1.

In the AISI ASD approach, no load factors are applied to the nominal loads to obtain the design loads in the evaluation of the design strengths. In the AISI LRFD approach, on the other hand, various linear combinations of the nominal loads are used for the design loads which are then compared with the design strengths. The design strength is the nominal strength divided by a factor of safety (Ω) in the AISI ASD approach, and the nominal strength multiplied by a resistance factor (ϕ) in the AISI LRFD approach. In the following paragraphs, the design constraints are formulated in a general form applicable to both AISI ASD and LRFD Specifications. The terms design loads and design strengths, however, depend on the design approach used.

3.3.1   Bending Strength Constraint

The maximum design bending moment, M_{max}, must be less than or equal to the design bending strength, M_d.

$$\frac{M_{max}}{M_d} - 1 \leq 0 \tag{3.5}$$

For beams with full lateral support, M_d is the minimum value obtained from considerations of initiation of yielding or inelastic reserve capacity and the shear lag effects. If the beam does not have full lateral support Eq. (3.5) has to be satisfied in each unbraced segment where M_{max} is the maximum design bending moment in the segment under consideration and M_d is the design lateral buckling strength of that segment.

Unlike hot-rolled shapes, cold-formed steel shapes are characterized by small wall thicknesses. As a result, elements under compression can buckle locally at stresses much lower than the yield stress. To approximate the non-uniform distribution of stresses in such compression elements the concept of an effective width is used (AISI, 1989 and 1991). In this approach, the lightly stressed portion of the element is assumed ineffective in resisting stresses. The stress distribution over the remaining portion is assumed to be constant.

Figure 3.2 shows the effective widths of cold-formed steel shapes in bending. The effective width b of the compression flange depends on the normal stress f in the flange, the flat-width-to-thickness ratio (X_1/X_4), and the plate buckling coefficient. Figures 3.2a and 3.2b represent shapes with stiffened and unstiffened compression flanges, respectively. The effective widths b_1 and b_2 of the web are a function of the maximum compressive and tensile stresses f_1 and f_2 (Figure 3.2) and the flat-width-to-thickness ratio (X_2/X_4).

Considering the local buckling effects, $M_d = S_e F_y / \Omega$ for ASD and $M_d = \phi_b S_e F_y$ for LRFD where S_e is the effective section modulus, $\Omega = 1.67$ is the factor of safety in bending, ϕ_b is the resistance factor for bending strength (equal to 0.90 for I, Z, and all other shapes not braced throughout, and 0.95 for fully braced hat shapes), and F_y is the yield stress of steel. The dependence of

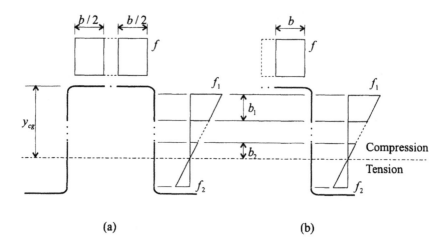

Figure 3.2   Effective widths of cold-formed steel beams in bending: (a) Shapes with stiffened compression flange; (b) Shapes with unstiffened compression flange

effective widths on stresses means that S_e (and other cross-sectional properties) cannot be calculated explicitly.

Following the AISI ASD and LRFD Specifications, the following stepwise procedure is used to determine the value of S_e at initiation of yielding in the cross-section:

Step 1: Assume the compression flange yields, i.e., $f = F_y$. Calculate b and the cross-sectional properties including the distance of the neutral axis from the outermost compression fiber, y_{cg}. If y_{cg} is greater than $d/2$ where d is the depth of the cross-section, go to step 2. If y_{cg} is less than $d/2$, skip step 2 and go to step 3.

Step 2: The assumption $f = F_y$ is correct. Calculate widths b_1 and b_2 based on section properties calculated in step 1. If $b_1 + b_2 \geq h_c$ where h_c is the flat height of the compression region of the web, stop. The web is fully effective as assumed and the calculated properties are correct. Otherwise, set the iteration number $n = 1$ and do the following:

1. Set $y_{cg}^n = y_{cg}$.

2. Calculate the cross-sectional properties with effective widths b, b_1, and b_2.

3. If $y_{cg} - y_{cg}^n < \varepsilon$ where ε is the stopping tolerance, stop. The current properties are the correct values within the tolerance.

4. Update b_1 and b_2 (b remains unchanged as $y_{cg} > d/2$ and $f = F_y$).

5. $n = n+1$.

6. Go to 1.

Step 3: The assumption that yielding initiates in the compression flange is not correct. Set the iteration number $n = 1$ and do the following:

1. Set $y_{cg}^n = y_{cg}$.

2. Calculate f with neutral axis at y_{cg}^n and the maximum stress in the tension flange at F_y. Calculate b.

3. Determine cross-sectional properties with effective width b and assuming the web is fully effective.

4. If $y_{cg}^n - y_{cg} < \varepsilon$ stop. Calculate b_1 and b_2. If $b_1 + b_2 \geq h_c$ the assumption that the web is fully effective is correct and the calculated properties are the desired effective cross-sectional properties. Otherwise, set iteration counter $m = 0$ and do the following:

a) Set $y_{cg}^m = y_{cg}$.

b) Calculate the cross-sectional properties with effective widths b, b_1, and b_2.

c) If $y_{cg} - y_{cg}^m < \varepsilon$ stop. The current properties are the correct values within the tolerance.

d) Update b, b_1, and b_2.

e) $m = m + 1$.

f) Go to a.

5. Update f and b.

6. $n = n + 1$.

7. Go to 1.

Some cold-formed steel shapes may be sufficiently compact to develop partial plastification. The AISI ASD and the LRFD Specifications recognize this inelastic reserve capacity and allow designers to increase the bending capacity by 25 percent provided certain provisions are satisfied. Our computational model takes into account this increase whenever the conditions are satisfied.

For short span beams with unusually wide flanges supporting concentrated loads the effective widths of both compression and tension flanges may be limited by shear lag effect. Shear lag causes the normal flexural stresses in the flanges to decrease with increasing distance from the web. As the span-to-flange width ratio increases the effective width of the flange decreases. For beams in which shear lag is important the design bending strength is the minimum of the values computed from local buckling considerations and shear lag effects.

The bending strength of laterally unbraced segments depends on the length of the segment, the variation of bending moment over the segment, the depth of the shape (d), and the moment of inertia of the compression portion of the shape with respect to the minor axis of bending. Note that the hat shape is laterally stable when bending with the top flange in compression.

3.3.2   Shear Strength Constraint

The maximum design shear, V_{max}, must be less than or equal to the design shear strength, V_d.

$$\frac{V_{max}}{V_d} - 1 \leq 0 \tag{3.6}$$

Yielding and local shear buckling of the web is controlled by this constraint. The appropriate relationship for V_d depends on the flat height (X_2), and the thickness (X_4) of the web.

3.3.3   Constraint on Combined Bending and Shear Strength

Bending and shear do not occur in isolation but rather interact with each other. The AISI ASD and the LRFD Specifications require that the following nonlinear interaction equation be satisfied throughout the beam:

$$\left(\frac{M_x}{M_d}\right)^2 + \left(\frac{V_x}{V_d}\right)^2 - 1.0 \leq 0 \tag{3.7}$$

where M_x and V_x are the design bending moment and shear at the location x, respectively. To find the location at which the constraint is most violated we evaluate the above expression at a number of equally spaced locations (including locations of concentrated loads) and include the one with the largest absolute value in the constraint set.

3.3.4   Constraint on Web Crippling Strength

In cold-formed steel construction, it is usually impractical to provide stiffeners at reactions and locations of concentrated loads. Because of the large flat-width-to-thickness ratios of webs, crippling is an important consideration in the design of cold-formed steel members. To prevent web crippling the following constraints must be satisfied:

$$\frac{R_A}{P_d} - 1 \le 0 \tag{3.8}$$

$$\frac{R_B}{P_d} - 1 \le 0 \tag{3.9}$$

$$\frac{P_i}{P_d} - 1 \le 0 \quad i = 1, N \tag{3.10}$$

where R_A and R_B are the left and right beam reactions, respectively, and P_i is the design concentrated load i located at a distance x_i from the left support. The design web crippling strength, P_d, is given by one of four formulas for a particular shape. The appropriate equation to be used depends on: (1) the distance of the concentrated load from the beam support, and (2) the spacing between the concentrated load and the nearest concentrated load in the opposite direction. All distances are measured from the edge of the bearing plates whose width, B, is a parameter in the formulas.

3.3.5 Constraint on Combined Web Crippling and Bending Strength

The combined effect of concentrated load and bending is accounted for by the following constraint:

$$P\left(\frac{P_i}{P_d}\right) + Q\left(\frac{M_i}{M_d}\right) - 1.0 \le 0 \quad i = 1, N; N \ne 0 \qquad (3.11)$$

where M_i is the design moment at the location of concentrated load i, and P and Q are constants depending on the shape of cross section and type of specification. For example, for I shape beams designed according to the AISI LRFD Specification P and Q are equal to 0.6212 and 0.7575, respectively.

3.3.6 Deflection Constraint

Maximum deflection, Δ_{max}, of the beam must be less than or equal to the allowable deflection, Δ_a.

$$\left(\frac{\Delta_{max}}{\Delta_a}\right) - 1.0 \le 0 \qquad (3.12)$$

The maximum deflection is determined at service loads. Because the effective widths of the compression elements depend on flexural stresses the moment of inertia varies with the bending moment along the length of the beam. The effective moment of inertia used is that at maximum bending moment. This results in a small error which is on the conservative side (Yu, 1991). The effective cross-sectional properties for deflection determination are calculated by a procedure similar to that used for bending strength determination.

3.3.7   Constraint on Flange Curling

Curling of both tension and compression flanges towards the neutral axis is a concern for unusually wide and thin flanges. Excessive curling reduces the bending capacity of the section and impairs its appearance. To limit flange curling to a given value, c_f, the following constraints must be satisfied:

$$\frac{\dfrac{X_1}{N_w} + r}{w_f} - 1.0 \leq 0 \qquad\qquad (3.13)$$

$$\frac{\dfrac{X_3}{N_w} + r}{w_f} - 1.0 \leq 0 \qquad\qquad (3.14)$$

where w_f represents the limiting value for the length of the flange projecting beyond the web for unstiffened flanges (the I and Z shapes) or half the distance between webs for stiffened flanges (the hat shape) and is given by

$$w_f = \left(\frac{0.061 dX_4 E}{f_{av}} \right)^{\frac{1}{2}} \left(\frac{100 c_f}{d} \right)^{\frac{1}{4}} \tag{3.15}$$

in which f_{av} is the average stress in the full, unreduced flange width. Curling on the order of 5 percent of the beam depth is not uncommon. Appearance considerations may dictate the choice of c_f.

3.3.8 Local Buckling Constraints

The AISI ASD and the LRFD Specifications require that the flat-width-to-thickness ratio of the compression flange be limited to 500 for stiffened flanges and 60 for unstiffened flanges. For beams with no web stiffeners, which is often the case, the web flat-width-to-thickness ratio is limited to 200.

For flange local buckling, when the top flange is in compression:

$$\frac{X_1 / X_4}{500} - 1.0 \leq 0 \quad \text{for hat shape} \tag{3.16}$$

$$\frac{X_1 / X_4}{60} - 1.0 \leq 0 \quad \text{for I and Z shapes} \tag{3.17}$$

and when the bottom flange is in compression:

$$\frac{X_3 / X_4}{60} - 1.0 \leq 0 \quad \text{for hat, I, and Z shapes} \tag{3.18}$$

For web local buckling:

$$\frac{X_3 / X_4}{200} - 1.0 \leq 0 \quad \text{for hat, I, and Z shapes} \tag{3.19}$$

In addition to the above constraints, practical considerations may dictate some upper and lower bounds on the design variables. For example, the thickness of a shape may be limited by the capacity of the forming equipment. Generally, the thickness of commonly used members varies from 0.5 mm to 6.0 mm.

3.4   NEURAL DYNAMICS OPTIMIZATION MODEL

Adeli and Park (1998) present a general neural dynamics model for optimization problems that guarantees a stable and local optimum solution. This solved one of the fundamental problems of using ANNs for optimization: How to find a neural dynamics system for a particular optimization problem that would produce a stable and local optimum solution. The neural dynamics optimization model is robust and particularly effective for large and complex optimization problems.

The constrained optimization problem (Eqs. 3.1–3.3), can be written in the following reduced form by combining Eqs. (3.2) and (3.3):

Minimize

$$W = F(\mathbf{X}) \tag{3.20}$$

subject to inequality constraints

$$g_j^k(\mathbf{X}) \le 0 \quad j = 1, J; k = 1, N_L \tag{3.21}$$

where J is the total number of inequality constraints including side constraints represented by Eq. (3.3). Using the exterior penalty function method, the constrained optimization problem can be written as an unconstrained optimization problem as follows:

$$V(\mathbf{X}, r_n) = F(\mathbf{X}) + \frac{r_n}{2} \left\{ \sum_{j=1}^{J} \left[g_j^+(\mathbf{X}) \right]^2 \right\} \qquad (3.22)$$

where $g_j^+(\mathbf{X}) = \max\{0, g_j(\mathbf{X})\}$ and r_n is the penalty parameter magnifying constraint violations.

A dynamic system is defined as a system of first order differential equations of the state variables:

$$\frac{d\mathbf{X}}{dt} = \dot{\mathbf{X}} = f(\mathbf{X}) \qquad (3.23)$$

where the vector $\mathbf{X}(t) = \{X_1(t), X_2(t), ..., X_N(t)\}^T$ represents the time evolution of node activations. A solution $\hat{\mathbf{X}}$ to the system of equations $\dot{\mathbf{X}} = 0$ is an equilibrium point of the system. In the theory of nonlinear system dynamics, the Lyapunov stability theorem is used to determine the stability of the equilibrium point. The theorem states that $\hat{\mathbf{X}}$ is stable if the time derivative of a functional, V, called the Lyapunov functional (or energy functional), is negative semi-definite for all non-zero $\mathbf{X}$, i.e. $\frac{dV}{dt} \leq 0$. The Lyapunov functional is defined as an analytic function of the system variables, $V(\mathbf{X})$, such that $V(\mathbf{0}) = 0$ and $V(\mathbf{X}) > 0$ for all $|\mathbf{X}| > 0$ (Kolk and Lerman, 1992).

In the design of cold-formed steel beams, each term X_i in the vector $\mathbf{X}$ represents a cross-sectional dimension and is always greater than zero. Therefore, both the objective function and the penalized constraint functions satisfy the conditions for a Lyapunov functional. As Eq. (3.22) is made of these two terms it is therefore a valid Lyapunov functional.

Using the chain rule, we have

$$\frac{dV}{dt} = \left(\frac{dV}{d\mathbf{X}}\right)\left(\frac{d\mathbf{X}}{dt}\right) = \left[\nabla F(\mathbf{X}) + r_n\left\{\sum_{j=1}^{J} g_j^+ \nabla g_j(\mathbf{X})\right\}\right]\dot{\mathbf{X}} \qquad (3.24)$$

where $\nabla F(\mathbf{X})$ and $\nabla g_j(\mathbf{X})$ are the gradients of the objective function and the jth constraint function, respectively. To satisfy the Lyapunov stability theorem, following Adeli and Park (1995b) we define the neural dynamics system as:

$$\frac{d\mathbf{X}}{dt} = \dot{\mathbf{X}} = -\nabla F(\mathbf{X}) - r_n\left\{\sum_{j=1}^{J} g_j^+ \nabla g_j(\mathbf{X})\right\} \qquad (3.25)$$

In this case

$$\frac{dV}{dt} = -\left[\nabla F(\mathbf{X}) + r_n\left\{\sum_{j=1}^{J} g_j^+ \nabla g_j(\mathbf{X})\right\}\right]^2 \leq 0 \qquad (3.26)$$

and the Lyapunov stability theorem is satisfied. This ensures that the system evolves such that the value of the penalized objective function always decreases.

To guarantee that an equilibrium point $\hat{\mathbf{X}}$ is an optimal solution of the problem the Kuhn-Tucker optimality conditions are satisfied (Adeli and Park, 1995b):

$$\frac{\partial L}{\partial X_i} = \frac{\partial F(\mathbf{X})}{\partial X_i} + \sum_{j=1}^{J} u_j \frac{\partial g_j(\mathbf{X})}{\partial X_i} = 0; \quad \text{for all } i \tag{3.27}$$

$$g_j(\mathbf{X}) + s_j^2 = 0; \quad j = 1, J \tag{3.28}$$

$$u_j s_j = 0; \quad j = 1, J \tag{3.29}$$

$$u_j \geq 0; \quad j = 1, J \tag{3.30}$$

where L is the Lagrangian function defined as a linear combination of the objective and constraint functions:

$$L(\mathbf{X}, \mathbf{u}, \mathbf{s}) = F(\mathbf{X}) + \sum_{j=1}^{J} u_j \left[g_j(\mathbf{X}) + s_j^2 \right], \tag{3.31}$$

in which s_j is the slack term for the jth inequality constraint, and u_j is the Lagrangian multiplier corresponding to the jth inequality constraint.

The optimum solution can be found by the integration:

$$\mathbf{X} = \int \dot{\mathbf{X}} dt \tag{3.32}$$

This integration is done by either the Euler or Runge-Kutta method. For this neural dynamics model the relationship, $\dot{\mathbf{X}} = 0$, is the learning rule which governs the evolution of the network towards a local optimal solution of the optimization problem.

3.5   NEURAL DYNAMICS MODEL FOR OPTIMIZATION OF COLD-FORMED STEEL BEAMS

The neural dynamics model for optimization of cold-formed steel beams has two major components: (1) An information server composed of modules representing the cold-formed steel beam design problem, and (2) a neural network topology representing the dynamic system corresponding to the optimization problem. The neural network topology, the modules in the information server, and the interaction between the two are shown in Figure 3.3.

The purpose of the information server is to obtain data from the user, define the cold-formed steel beam design problem, and to provide the necessary information to the neural dynamics model. The user input is obtained only once in the beginning. Information to the neural dynamics model, on the other hand, is provided at each iteration based on the updated values of the design variables. The information server has three modules that interact with one another. The *Shapes* module provides information on the selected shape to the other modules and the neural dynamics model. It also computes the objective function (weight of the beam) and its gradients. The design loads (bending moments and shear) are computed by the *Beam analysis* module. This module is used only once in the beginning as the span, the bracing conditions, and the design loads are constants in the optimization problem. The *AISI Specifications* module contains the provisions of both AISI ASD and LRFD Specifications. At each

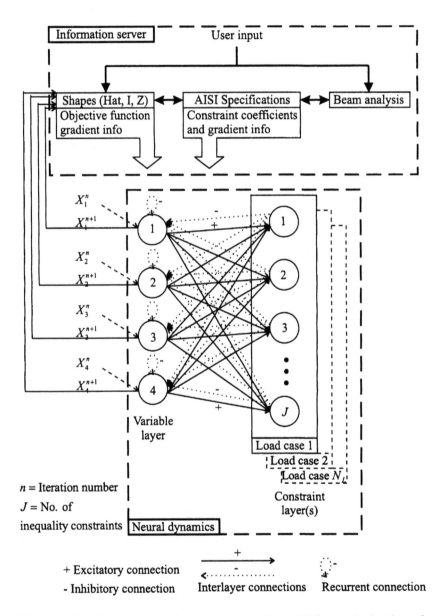

Figure 3.3   Components of neural network model for optimization of cold-formed steel beams

iteration it provides the coefficients of the constraint functions and their gradients to the neural dynamics model.

In general, the topology of a neural network is represented by a matrix of weighted connections between vectors (or layers) of nodes. The input to a node in a layer is the weighted sum of outputs of nodes in the connected layer(s). The output of a node is obtained by applying an appropriate activation function to the input. The operation of the network is governed by a learning rule that controls the evolution of the connection weights or the node outputs. This learning rule needs to converge to a stable state representing the desired solution.

The topology of the neural dynamics model for optimization of cold-formed steel beams has one variable layer and N_L constraint layers. The variable layer has four (4) nodes corresponding to the design variables in the design of cold-formed steel beams, that is, the cross-sectional dimensions identified in Figure 3.1. The nodes in each constraint layer correspond to the design constraints for a particular loading case. All constraint layers are fully interconnected with the variable layer. Both excitatory (having positive weights) and inhibitory (having negative weights) connections are used to link the nodes. In addition to the commonly used interlayer connections, recurrent connections are also used in the variable layer.

The neural dynamics algorithm for optimization of cold-formed steel beams follows:

Step 1: Set the iteration counter, $n = 0$.

Step 2: Select an initial decision vector $\mathbf{X}^n = \mathbf{X}^0$, the initial penalty parameter r_0, and the objective function tolerance, δ, used in the stopping criterion.

Step 3: Calculate the gradients of the objective function and assign them to the inhibitory recurrent connections of the variable layer. For the ith node in the variable layer, the weight of the recurrent connection is:

$$C_i = -\frac{\partial F(\mathbf{X}^n)}{\partial X_i} \quad i = 1,4 \tag{3.33}$$

C_i represents the direction of steepest descent of the objective function along X_i.

Step 4: Assign the coefficients of the constraint functions to the excitatory connections from the variable layer to the constraint layer(s). The input to the jth node in the kth constraint layer is therefore the magnitude of the constraint, $g_j^k(\mathbf{X}^n)$.

Step 5: Calculate the output of the nodes in the constraint layer(s). The output of node j in layer k is given by:

$$O_{cj}^k = r_n \max\left[0, g_j^k(\mathbf{X}^n)\right] \quad j = 1, J; k = 1, N_L \tag{3.34}$$

This is the activation function of the node. The output is zero when the constraint is satisfied and equal to the penalized constraint violation otherwise. If there is only one load case skip step 6 and go to step 7.

Step 6: Create a competition between the outputs of the corresponding nodes in the various constraint layers. The governing output of the jth node is taken as the maximum value obtained from all the loading cases.

$$O_{cj} = \max\left[O_{cj}^1, O_{cj}^2, O_{cj}^3, \ldots, O_{cj}^{N_L}\right] \quad j = 1, J \tag{3.35}$$

In this way, only the most violated constraint is included in the constraint set.

Step 7: Assign the gradients of the violated constraint functions to the inhibitory connections from the constraint layer to the variable layer. The weight of the connection from the jth constraint node to the ith variable node is given by:

$$Z_{ji} = \frac{\partial g_j(\mathbf{X}^n)}{\partial X_i} \quad j = 1, J; i = 1,4 \tag{3.36}$$

All gradients are calculated by the finite difference method.

Step 8: Calculate the input to the variable layer. For the ith node in the variable layer, the input is given by:

$$I_{vi} = C_i + \sum_{j=1}^{J} Z_{ji} O_j \quad i = 1,4 \tag{3.37}$$

I_{vi} is the direction of the steepest descent of the equivalent unconstrained optimization problem that reduces X_i.

Step 9: Update the decision variables using the following learning rule:

$$X_i^{n+1} = X_i^n + \int I_{vi} dt \quad i = 1,4 \qquad (3.38)$$

The Euler method is used to evaluate the integral.

Step 10: Calculate the new value of the objective function, $F(\mathbf{X}^{n+1})$.

If

$$\frac{|F(\mathbf{X}^n) - F(\mathbf{X}^{n+1})|}{F(\mathbf{X}^n)} < \delta \qquad (3.39)$$

stop. The current state vector, $\mathbf{X}^{n+1}$, is the optimal solution of the problem. Otherwise, set $n = n + 1$ and update the penalty parameter using the expression (Adeli and Park, 1995b):

$$r_n = r_o + \frac{n}{\alpha} \qquad (3.40)$$

where α is a real positive number. This function is chosen to avoid the possibility of numerical ill-conditioning by gradually increasing the penalty with increasing iterations. Go to step 3.

3.6   APPLICATION OF THE MODEL

Three simply-supported example beams are used to test the developed neural network model for optimization of cold-formed steel beams. They include hat, I, and Z-shaped beams designed according to the AISI ASD and the LRFD Specifications subjected to various loading conditions and having different lateral bracing conditions. The following data have been used in all the examples: $\rho =$ Unit weight of steel = 77 MN/mm^3 (0.2836 lb/in^3); $F_y =$ Yield stress = 345 N/mm^2 (50.8 ksi); E = Modulus of elasticity = 203 kN/mm^2 (29.44 × 10^3 ksi); Δ_a = Allowable maximum deflection = $L/240$; B = Length of bearing plate at reactions and locations of concentrated loads = 90 mm (3.54 in.); c_f = Allowable amount of flange curling = 6 mm (0.24 in.); r = Inner radius at corners of shape = 8 mm (0.31 in.); X_4^U = 8 mm (0.31 in.). No lower limit is placed on the thickness of the shapes. For the stopping criterion, $\delta = 1 \times 10^{-4}$, and $\varepsilon = 1 \times 10^{-3}$.

3.6.1 Example 1

In this example a simply-supported beam of span length L = 3 m is subjected to uniformly distributed dead and live loads of 3 kN/m and 12 kN/m, respectively. Full lateral support is assumed for all the cases. Three types of shapes are designed for this beam: hat, I, and Z shapes. Two sets of initial design variables are used in order to investigate the effects of the initial design on the optimum solution: A{50, 50, 150, 8 mm} and B{100, 100, 200, 8 mm}.

The optimization results are summarized in Table 3.1 and convergence results are shown in Figure 3.4. A number of important observations can be made. First, the dominant design variable is the thickness. For the optimum solution it can change drastically from the initial design value. Second, the optimization algorithm yields a local optimum solution in the vicinity of the initial values for the depth and the flange widths. Third, in all the examples the bending strength and the combined bending and shear strength control the

Table 3.1 Optimum solutions for Example 1

Case #	Shape	Specification	X_1 mm (in.)	X_2 mm (in.)	X_3 mm (in.)	X_4 mm (in.)	W N (lb)
Initial Design A							
1	Hat	ASD	20 (0.8)	20 (0.8)	100 (3.9)	6.0 (0.24)	611.3 (137.4)
2	Hat	LRFD	31 (1.2)	31 (1.2)	119 (4.7)	3.0 (0.12)	347.0 (78.0)
3	I	ASD	31 (1.2)	31 (1.2)	119 (4.7)	3.0 (0.12)	347.0 (78.0)
4	I	LRFD	31 (1.2)	31 (1.2)	119 (4.7)	3.0 (0.12)	347.0 (78.0)
5	Z	ASD	40 (1.6)	40 (1.6)	132 (5.2)	3.1 (0.12)	221.1 (49.7)
6	Z	LRFD	40 (1.6)	40 (1.6)	132 (5.2)	3.1 (0.12)	221.1 (49.7)
Initial Design B							
1	Hat	ASD	20 (0.8)	20 (0.8)	150 (5.9)	6.0 (0.24)	749.9 (168.6)
2	Hat	LRFD	36 (1.4)	36 (1.4)	143 (5.6)	2.1 (0.08)	266.6 (59.9)
3	I	ASD	36 (1.4)	36 (1.4)	143 (5.6)	2.0 (0.08)	258.5 (58.1)
4	I	LRFD	36 (1.4)	36 (1.4)	143 (5.6)	2.1 (0.08)	266.6 (59.9)
5	Z	ASD	36 (1.4)	36 (1.4)	143 (5.6)	2.1 (0.08)	272.3 (61.2)
6	Z	LRFD	36 (1.4)	36 (1.4)	143 (5.6)	2.8 (0.11)	203.2 (45.7)
			36 (1.4)	36 (1.4)	143 (5.6)	2.8 (0.11)	203.2 (45.7)

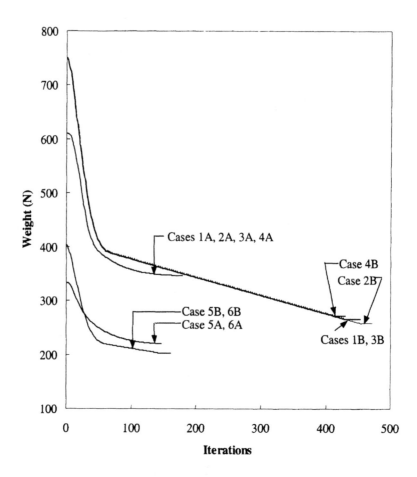

Figure 3.4   Convergence histories for Example 1

optimum design. For the live-to-dead load ratio of 4 used in this example, there is no significant difference in the optimum designs based on the AISI ASD and the LRFD Specifications.

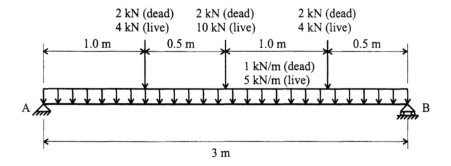

Figure 3.5   Example 2

3.6.2   Example 2

In this example, the beam is subjected to concentrated loads in addition to uniformly distributed load, as shown in Figure 3.5. Three different lateral support conditions are considered as noted in Table 3.2. For this example an I shape is optimized using both AISI ASD and LRFD Specifications. Again, all cases are solved using two initial sets of design variables: A{70, 70, 200, 8 mm} and B{70, 70, 300, 8 mm}.

The optimization results are summarized in Table 3.2 and the convergence results shown in Figure 3.6. For cases 1 and 2, the lateral buckling strength of the unbraced segment controls the optimum design. For cases 3 and 4, the lateral buckling strength of the unbraced segment(s), the bending strength, and the combined bending and shear strength are all active at the optimum design. The optimum designs for cases 5 and 6 are controlled by bending strength and the combined bending and shear strength. The other conclusions

Table 3.2   Optimum solutions for Example 2

Case #	Lateral bracing	Specification	X_1 mm (in.)	X_2 mm (in.)	X_3 Mm (in.)	X_4 mm (in.)	W N (lb)
Initial Design A							
1	Full	ASD	50 (2.0)	50 (2.0)	150 (5.9)	8.0 (0.31)	1110.8 (249.7)
2	Full	LRFD	72 (2.8)	72 (2.8)	199 (7.8)	3.0 (0.12)	518.4 (116.6)
3	Midspan	ASD	73 (2.9)	73 (2.9)	200 (7.9)	3.0 (0.12)	521.4 (117.2)
4	Midspan	LRFD	89 (3.5)	89 (3.5)	199 (7.8)	3.0 (0.12)	543.8 (122.3)
5	None	ASD	90 (3.5)	90 (3.5)	200 (7.9)	3.0 (0.12)	546.6 (122.9)
6	None	LRFD	131 (5.2)	131 (5.2).	208 (8.2)	3.0 (0.12)	620.2 (139.4)
			132 (5.2)	132 (5.2).	210 (8.3)	3.0 (0.12)	624.0 (140.3)
Initial Design B							
1	Full	ASD	50 (2.0)	50 (2.0)	250 (9.8)	8.0 (0.31)	1480.4 (332.8)
2	Full	LRFD	61 (2.4)	61 (2.4)	244 (9.6)	2.5 (0.10)	459.7 (103.3)
3	Midspan	ASD	61 (2.4)	61 (2.4)	244 (9.6)	2.5 (0.10)	464.3 (104.4)
4	Midspan	LRFD	64 (2.5)	64 (2.5)	246 (9.7)	3.0 (0.12)	570.1 (128.2)
5	None	ASD	65 (2.6)	65 (2.6)	246 (9.7)	3.0 (0.12)	571.3 (128.4)
6	None	LRFD	109 (4.3).	109 (4.3).	260 (10.2)	3.1 (0.12)	662.3 (148.9)
			111 (4.4)	111 (4.4).	261 (10.3)	3.0 (0.12)	655.8 (147.4)

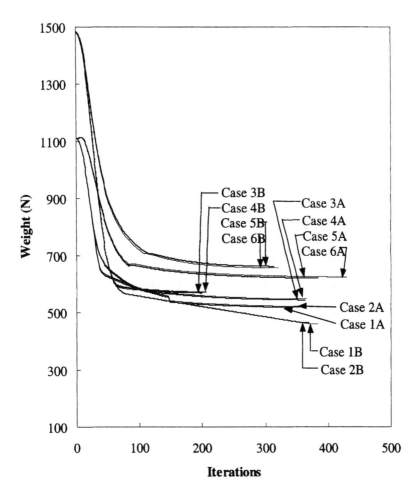

Figure 3.6   Convergence histories for Example 2

regarding the thickness and the local optimum solutions reached for
Example 1 are also valid for this example.

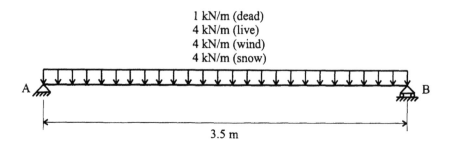

Figure 3.7   Example 3

3.6.3   Example 3

In this example a 4 m long Z-shaped roof purlin is designed (Figure 3.7). The loading consists of uniformly distributed dead, live, snow, and wind loads. Different bracing conditions are considered as indicated in Table 3.3. All cases are solved starting from two sets of initial design variables: A{100, 100, 200, 8 mm} and B{70, 70, 300, 8 mm}.

The optimum solutions and the convergence results are given in Table 3.3 and Figure 3.8. For cases 1 and 2, the lateral buckling strength of the unbraced segment controls the optimum design. For cases 3A to 6A, all three of the following constraints are active at the optimum design: lateral buckling strength of the unbraced segments, the bending strength, and the combined bending and shear strength. The optimum designs for cases 7A and 8A are controlled by the bending and combined bending and shear strengths. Providing lateral bracing at the midspan reduces weight significantly as compared to the unbraced beam. However, additional lateral bracing has no

Table 3.3 Optimum solutions for Example 3

Case #	Lateral bracing	Specification	X_1 mm (in.)	X_2 mm (in.)	X_3 mm (in.)	X_4 mm (in.)	W N (lb)
Initial Design A							
1	Full	ASD	50 (2.0)	50 (2.0)	200 (7.9)	8.0 (0.31)	863.5 (194.1)
2	Full	LRFD	50 (2.0)	50 (2.0)	200 (7.9)	2.9 (0.11)	310.9 (69.9)
3	QP, midspan	ASD	50 (2.0)	50 (2.0)	200 (7.9)	3.0 (0.12)	328.9 (73.9)
4	QP, midspan	LRFD	50 (2.0)	50 (2.0)	200 (7.9)	2.9 (0.11)	310.9 (69.9)
5	Midspan	ASD	61 (2.4)	61 (2.4)	207 (8.2)	3.0 (0.12)	328.9 (73.9)
6	Midspan	LRFD	64 (2.5)	63 (2.5)	210 (8.3)	3.0 (0.12)	351.0 (78.9)
7	None	ASD	97 (3.8)	97 (3.8)	218 (8.6)	3.1 (0.12)	360.8 (81.1)
8	None	LRFD	100 (3.9)	99 (3.9)	220 (8.7)	3.1 (0.12)	421.8 (94.8)
							431.2 (96.9)
Initial Design B							
1	Full	ASD	50 (2.0)	50 (2.0)	250 (9.8)	8.0 (0.31)	971.3 (218.4)
2	Full	LRFD	61 (2.4)	61 (2.4)	245 (9.6)	2.9 (0.11)	366.3 (82.3)
3	QP, midspan	ASD	61 (2.4)	61 (2.4)	245 (9.6)	3.1 (0.12)	387.4 (87.1)
4	QP, midspan	LRFD	61 (2.4)	61 (2.4)	245 (9.6)	2.9 (0.11)	366.3 (82.3)
5	Midspan	ASD	61 (2.4)	61 (2.4)	245 (9.6)	3.1 (0.12)	387.4 (87.1)
6	Midspan	LRFD	61 (2.4)	61 (2.4)	245 (9.6)	2.9 (0.11)	366.3 (82.3)
7	None	ASD	89 (3.5)	89 (3.5)	253 (10.0)	3.0 (0.12)	432.1 (97.1)
8	None	LRFD	90 (3.5)	88 (3.5)	254 (10.0)	3.1 (0.12)	442.5 (99.5)

QP = Quarter points.

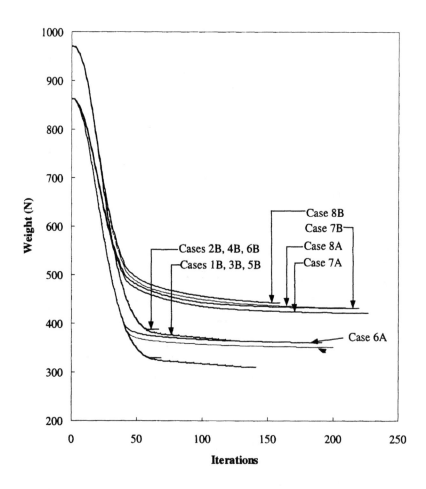

Figure 3.8   Convergence histories for Example 3

significant effect on the optimum designs. The web crippling strength at the reactions controls the optimum design of cases 3B to 8B. As a result, for these cases the lateral bracing condition has no effect on the optimum design. The same conclusions are reached regarding the thickness and the convergence to a local optimum solution as for examples 1 and 2.

3.7   GLOBAL OPTIMUM DESIGN CURVES FOR HAT-SHAPED BEAMS

As stated earlier, cold-formed steel members have a major advantage over hot-rolled steel shapes: they can be easily shaped and sized to meet any particular design requirement. As such, they provide a much larger variety of choices for steel designers. The result often is a lighter and more economical section compared with hot-rolled steel beams for low-rise building structures when the beam spans are not long. The price to be paid for this versatility is the complicated iterative design process. And finding the optimum or minimum weight beam is a challenging problem considering the complex and highly nonlinear constraints of the AISI ASD and LRFD Specifications that govern their design.

To demonstrate the robustness and practical applicability of the neural dynamics model, we perform an extensive parametric study and develop global optimum design curves for hat-shaped cold-formed steel beams as an example. The basis of design is the AISI ASD Specification (AISI, 1989). The loading on the beam is assumed to be a uniformly-distributed load of intensity q (Figure 3.1a). The variables in the parametric study are the span length (L), the load intensity (q), the yield stress of steel (F_y), and the lateral bracing condition. For each set of cross-sectional shape and yield stress of steel the global optimum values of the thickness ($t = X_4$), the web flat-depth-to-thickness ratio ($d/t = X_2/X_4$), and the flange flat-width-to-thickness ratio ($b/t = X_1/X_4 = X_3/X_4$) are obtained. They are plotted as a function of the span length.

In the parametric studies, span length is varied from 2 m to 8 m. The load intensity is varied from 2.5 kN/m to 20 kN/m. Metric values of material properties are used from the ASTM Standards (ASTM, 1996). Design curves are presented for two different grades of steel with yield stress of 250 N/mm^2 (for ASTM A36M) and 345 N/mm^2 (for ASTM A570M). The top and bottom flange widths are constrained to have equal dimensions (i.e. $X_1 = X_3$). Values of other parameters chosen are given in Section 3.6.

3.7.1   Parametric Studies and Search for Global Optima

A local optimum is guaranteed whenever the Kuhn-Tucker optimality conditions are satisfied. A global optimum can be guaranteed only when both the objective function and the feasible design space are convex. Proving the convexity for large and complicated optimization problems is impractical. Practically all structural optimization problems for designs based on actual commonly-used design codes are non-convex (Adeli, 1994). In optimum design of beams according to the AISI ASD and LRFD Specifications, due to highly nonlinear and implicit nature of design constraints, numerous local optima exist and finding the global optimum is particularly a challenging problem.

If all constraints are active at a local optimum point, that point will in fact be the global optimum. But, in practical optimization problems this is almost never the case. In order to find the global optimum solution as quickly as possible we try to move towards the solution (local optimum) which has the maximum number of active constraints. In our parametric investigation of cold-formed beams, we found that the following constraints are most often the critical ones: bending strength, combined bending and shear strength, web crippling strength, deflection, and lateral buckling strength (for unbraced beams only).

The initial design specified in the optimization process may be feasible or infeasible. If a feasible initial design is chosen, our extensive parametric study indicates that the optimum value for the thickness is often drastically different from the initial value while the optimum values for the other parameters (b and d) remain close to the initial values. In other words, the dominant design variable is the thickness, and the result is a local optimum in the vicinity of the initial b and d values. On the other hand, if an infeasible initial design is chosen all design parameters (b, d, and t) change without one dominating the others, as the solution approaches a local optimum. Consequently, starting from an infeasible solution helps to move toward the global optimum faster.

In the parametric studies, an upper limit is placed on the thickness (t_{max}). We start from a large value and decrease it in decreasing increments. For each t_{max} various local optima are found for various ratios of b/t and d/t. These ratios are changed after observing their pattern of change and which constraints are active. We discovered that the global optimum changes little with the change in the value of the flange width, b. Consequently, a small initial value of b is chosen for various shapes. Only the values of t_{max} and d are changed in the parametric study. The following strategy is used for the search for the global optimum:

1. Set a large value for t_{max}.

2. Set the initial value for b to a small value (e.g. 10 mm) and t to a large value such that $t > t_{max}$.

3. Choose an initial value for d.

4. Run the algorithm. If a local optimum is found go to step 6. Otherwise, go to step 5.

5. Reduce t_{max}. Go to step 3.

6. Increase or decrease the initial value of d depending on which constraints are active and using heuristic rules (for example, d is increased when the bending strength constraint is active and it is decreased when the web crippling strength constraint is active). Go to step 3.

Following this strategy, the local optimum value is often reduced in subsequent trials with occasional fluctuations. This is a heuristic trial-and-error strategy that takes advantage of the behavior of cold-formed steel beams in order to reduce the huge search space for finding the global optimum. Through thousands of computer runs we discovered insights into the nature of the problems which in turn were used to expedite the search process. This point will be elaborated further in the next section.

Whenever two or more constraints are active at a local optimum point, which is often the case, pinpointing the global optimum becomes more demanding. In such cases the search for the global optimum has to be carried out in smaller increments of design variables t_{max} and d while keeping track of the dominant constraint.

3.7.2   Design Curves for Hat-Shapes

Figures 3.9, 3.10 and 3.11 show the global optimum design curves for thickness, web depth-to-thickness ratio, and flange width-to-thickness ratio for hat-shapes for yield stress of 345 N/mm^2. Figures 3.12 to 3.14 show similar results for yield stress of 250 N/mm^2. The hat-shape is laterally stable when the wide flange is in compression. Therefore, only the unbraced condition is considered. The bending, combined bending and shear, and web crippling strength constraints control the global optimum design for most cases. Deflection constraint becomes active for large span lengths (greater than about 4 m) particularly for small applied loading (usually less than 7.5 kN/m). Deflection constraint is less important for the lower strength steel where the strength constraints dominate the global optimum design.

For higher level of loading (greater than 7.5 kN/m) and larger span length the bending strength dominates the global optimum which is attained at larger values of d/t (solid lines in Figure 3.10). The corresponding b/t values approach a constant (solid lines in

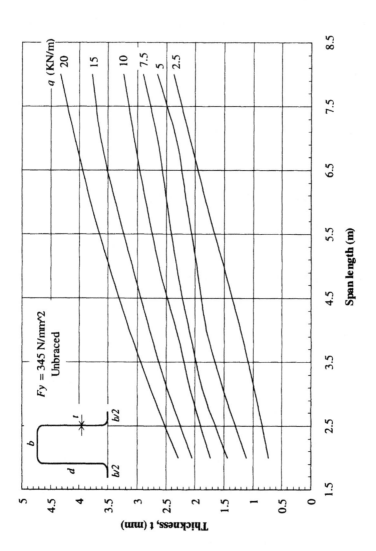

Figure 3.9 Plot of global optimum design thickness versus span length for unbraced hat-shape beams with $F_y = 345$ N/mm^2

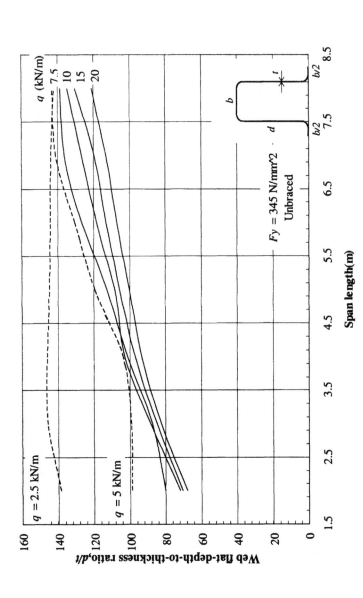

Figure 3.10 Plot of global optimum design web depth-to-thickness ratio versus span length for unbraced hat-shape beams with $F_y = 345$ N/mm^2

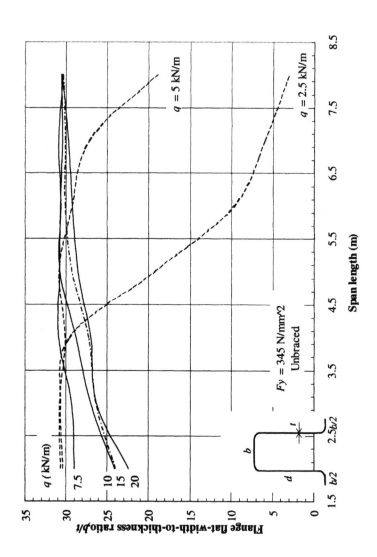

Figure 3.11 Plot of global optimum design flange width-to-thickness ratio versus span length for unbraced hat-shape beams with $F_y = 345$ N/mm²

Figure 3.11). This constant represents the maximum b/t ratio that will cause no local buckling in the compression flange. As such, when only the bending constraint is active the global optimum value for b/t is close to the boundary value between the fully effective and partially effective section obtained from the AISI ASD Specification (AISI, 1989) as follows:

$$\frac{b}{t} = 0.64 \sqrt{\frac{kE}{F_y}} \qquad (3.41)$$

where k is the plate buckling coefficient (equal to 0.43 for unstiffened flanges and 4 for stiffened flanges). For hat-shapes $k = 4$ and the optimum b/t value is equal to 31.0 for $F_y = 345$ N/mm^2 and 36.5 for $F_y = 250$ N/mm^2. Whenever the web crippling strength constraint is active at the global optimum, the value of d/t tends to decrease. Also, since the flanges have no contribution to the web crippling strength, the optimum values of b and b/t decrease with the dominance of web crippling constraint (solid lines in Figures 3.10 and 3.11 at shorter span lengths). Similar conclusions are made for the lower strength steel except that the strength constraints have a greater influence and deflection is not usually a problem.

Some interesting results for optimum values were discovered. Two curves in Figures 3.10 and 3.11 identified by dashed lines for lower intensities of $q = 2.5$ kN/m and $q = 5$ kN/m look interestingly different from the rest. For these curves, the global optimum values of b/t are much smaller for large span lengths and the global optimum values of d/t are much larger for smaller span lengths.

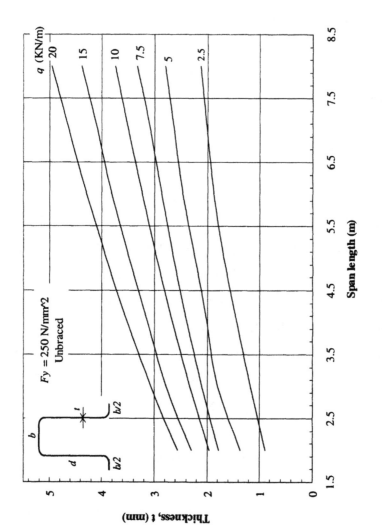

Figure 3.12 Plot of global optimum design thickness versus span length for unbraced hat-shape beams with F_y = 250 N/mm^2

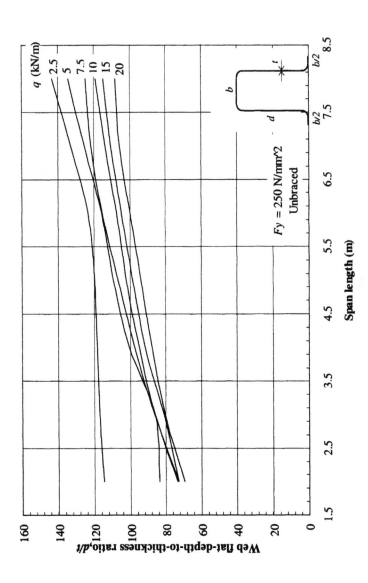

Figure 3.13 Plot of global optimum design web depth-to-thickness ratio versus span length for unbraced hat-shape beams with $F_y = 250$ N/mm²

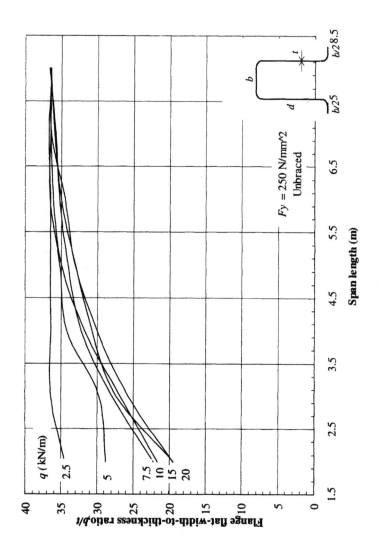

Figure 3.14 Plot of global optimum design flange width-to-thickness ratio versus span length for unbraced hat-shape beams with $F_y = 250$ N/mm²

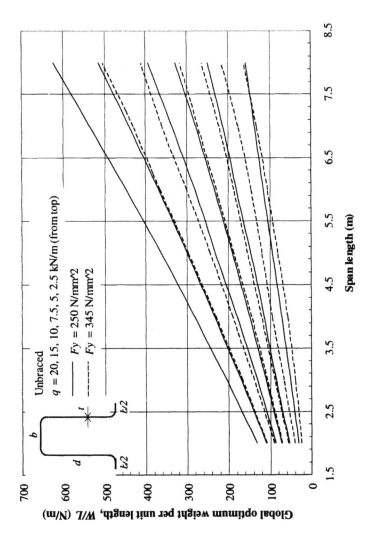

Figure 3.15 Comparison of global optimum weights per unit length for hat-shape beams with $F_y = 345$ N/mm^2 and $F_y = 250$ N/mm^2

These global optimum designs are controlled by the deflection and the web crippling strength constraints. Since the flanges have no contribution to the web crippling strength, the values of b and b/t tend to reduce. Also, since the deflection constraint is active the global optimum values of d/t tend to increase the stiffness of the section. These curves show the optimum trade-off between the deflection and the web crippling constraints.

Figure 3.15 shows the global optimum weight per unit length values for the two grades of steel used. The higher strength steel produces significantly lighter structure especially for 'heavier loads and longer spans. The weight of the global optimum beam made of the higher strength steel is 78–82 percent of the corresponding beam made of the lower strength steel.

3.8   CONCLUDING REMARKS

The design of cold-formed steel is governed by complex and highly nonlinear rules that make efficient optimization difficult. In this chapter, we presented a robust neural network model for optimum design of cold-formed steel beams and applied it to three commonly used shapes (hat, I, and Z) according to the AISI ASD and the LRFD Specifications. Any combination of uniformly distributed and/or concentrated loads can be modeled. The beam may be fully braced, unbraced, or braced at a specified number of points. Furthermore, we also demonstrated the practical applicability of the model by developing global optimum design curves for hat-shape beams.

The convergence curves obtained for a large number of examples including those presented in Figures 3.4, 3.6, and 3.8 demonstrate the robustness of the neural dynamics optimization model. Further, the computational model is efficient in terms of computer processing time. The optimum results for various examples presented in this chapter are obtained on an IBM RISC 6000 machine in less than 6.5 seconds.

This chapter introduced the neural dynamics model as a robust and practical tool for engineering decision making. In Chapter 5, we develop the new construction scheduling and cost optimization model for construction planning and management. In Chapter 6, the neural dynamics model is adapted for the solution of the problem.

4

PROJECT PLANNING AND MANAGEMENT AND CPM

4.1   INTRODUCTION

Construction projects are large and complicated undertakings with significant economic, political, and environmental ramifications. Typical civil engineering construction projects can range in cost from hundreds of thousands of dollars to billions of dollars. Moreover, large construction projects are often driven by political and environmental concerns. In any construction project, several participants or parties are involved from the project inception to its completion. These parties, whose stakes in the project may vary, often place varying importance on the economic, political, and environmental ramifications of the project.   Thus, a construction project is not only a complex engineering process but also a multiple-party goal satisfaction problem where each party desires the best solution.

An example of a complex construction project is a freeway expansion project wherein one additional lane is added. The need for such an expansion is usually determined from traffic studies done by the traffic agency. However, traffic studies alone may not dictate the final design or even the go-ahead for the project; other parties such as the tax paying public, the using public, the politicians, and the environmentalists can play significant roles in the project's conception and development.

The project design and contract specifications define the project's goals and requirements. Once this document is finalized a contractor is selected to carry out the required work for the project. The contractor and the traffic agency (owner or client) interact throughout the construction phase of the project to ensure that the project specifications are met to the satisfaction of both parties. A lot is at stake for both contractor and traffic agency during this phase as both parties have already invested significant resources into the project.

To ensure that the project goals are met to the satisfaction of the parties involved effective project planning and management is needed. Planning gives the project manager the opportunity to analyze the work required for the project beforehand and determine the most appropriate strategy based on resource availability, construction methodology, constructability, cost, and time. The result of this analysis is documented in a plan. Project management is needed to maintain the plan so that it correctly reflects the current state of the project and charts out the future strategy needed to achieve the project's goals.

An important task in project management is the evaluation of the plan. The owner has to verify the accuracy of the contractor's plan before it can be approved. Plans are also reviewed periodically to monitor progress and also when the contractor claims a change order. Change order claims are especially contentious as significant cost overruns and time delays may be involved. An objective way of analyzing such changes is essential to avoid any unfair advantage and the possibility of subsequent litigation.

A recent trend in construction project planning and management is to consider it in the context of concurrent and collaborative engineering (Adeli, 2000). Its effective solution requires not only sound engineering analysis and decision making, but also reliable information sharing and communication among the parties. This concept is shown schematically in Figure 4.1. The key components needed for effective project planning and management must be shared seamlessly by the contractor and the owner. In this book, we present new computational models for these components of project planning and management focusing on the engineering analysis and decision making needs of the owner and the contractor. We then present a software architecture that facilitates the integration of these models for concurrent and collaborative use. As shown in Figure 4.1 only the construction phase of the project's life cycle is considered. This phase, however, is the most important one in the overall project life cycle.

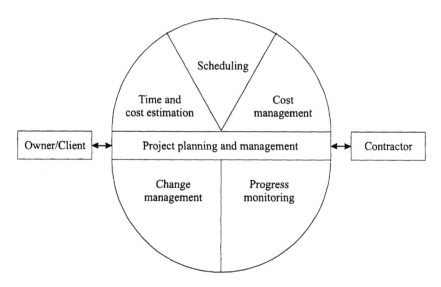

**Figure 4.1   Project planning and management as a concurrent and
collaborative engineering model**

In this chapter the basics of project planning and management are
introduced. The project is defined, and several attributes that
determine the state of a project at any given time are outlined.
Elements of project scheduling are described. In the latter part of the
chapter the network scheduling technique, and in particular the
critical path method (CPM) is presented as the current practice for
project scheduling.

4.2   WHAT IS A PROJECT?

4.2.1   Definition

A project is an endeavor to achieve a desired goal. The achievement
of this goal is not a trivial undertaking and thus requires careful

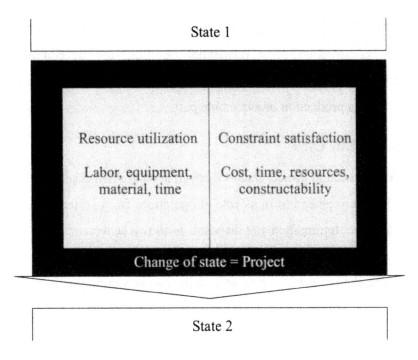

Figure 4.2   A project defined as a non-trivial change of state

planning and management. A project is made up of tasks, which can be thought of as mini-projects. These tasks, which collectively define the goal of the project, cannot be executed in any arbitrary order. Their execution sequence is constrained by limits on the duration and cost of the project, resource availability, construction methodology, and practical feasibility.

Figure 4.2 shows the definition of a project conceptually. A project is defined for every non-trivial state change from an existing state to a desired state or goal that consumes resources and requires constraint satisfaction. This definition is broad and encompasses projects of different scope and sizes. For example, a project can be the construction of a residential building, the development of engineering software, the organization of a technical conference, or the mass production of a machine part.

4.2.2   Life Cycle

A construction project passes through several distinct phases over its life. These phases form a cycle of operations from project conception to project termination and disposal, as shown schematically in Figure 4.3. A project is first conceived when a need is realized. At this stage, the project is just an idea in the owner's mind that is borne out of necessity. The idea is then developed in the feasibility study phase where project options and goals are evaluated in detail. It may happen that the project is found infeasible at this stage; in such a case the project is terminated immediately. Otherwise, work on the detailed design is undertaken. The result of this phase is the project specification, which outlines the project's goals and details the design and contract specifications. A project can pass directly from the conception phase to the design phase. This may be the case for low-cost and low-risk projects that do not involve many participants with conflicting goals.

In the construction and development phase, work is undertaken towards the achievement of the goals outlined in the project specification. This phase consumes resources at a rapid rate as the state of the project is changed towards the ultimate state or goal. As such, this phase is one of the most important phases in the project's

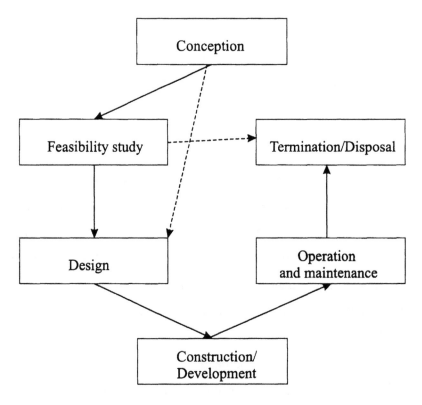

Figure 4.3   Project's life cycle

life cycle that has a direct bearing on the success or failure of the project. Once the facility is constructed the project goes into the operation and maintenance phase. And finally, when the useful life of the facility is complete the project is terminated with the decommissioning and/or destruction of the facility. Some projects, such as highway projects, may exist indefinitely through continuous maintenance and updating. In Figure 4.3 the phases are shown in a sequence one after the other without any feedback. In reality, this is usually not the case and two or more phases may co-exist at the same time. For example, the construction of a large civil engineering project may start once the design for the first stage is complete; the design for subsequent stages may continue concurrently with the construction.

4.2.3 Participants

Several participants (or parties) are associated with a project during its life cycle. The participants are either individuals or organizations that have a stake in the project and can influence the design, development, and final outcome of the project. The amount of influence a participant can exert depends on his/her/its stake (investment, risk, etc.) in the project, and therefore can vary from participant to participant.

There are two key participants in a construction project: the owner and the contractor. The owner conceives the project and is ultimately responsible for its outcome. As such, it has the largest stake in the project. The contractor is responsible for carrying out the work needed for the project in accordance with the project specification. The primary concern of the contractor is to do the job at minimum cost without violating any contract specification that can result in monetary and/or reputation loss. As the construction phase of a project involves large amounts of resource mobilization and utilization, this phase, and in particular, the relationship between the owner and the contractor during this phase plays a key role in the project's life cycle. The goals of these parties are not identical, and because much is at stake during this phase for both participants, disputes can arise that can become contentious. Furthermore, a change order initiated by one of these parties will likely be contested by the other. Therefore, managing the interaction between the owner and the contractor during the construction phase must be an integral and essential component of effective project management.

Other project participants may include designers/developers, financial institutions, tax paying and consuming public, politicians, and groups such as environmental and historical preservation organizations. A distinction must be made between a participant and the role that a participant plays. For example, a project manager may be the owner itself or it may be an external consultant. Similarly, the role of a scheduler—the person/organization responsible for the development and maintenance of the schedule—may be played by the project manger or a separate individual.

4.2.4 Attributes of a Project

Projects can be characterized by several attributes. These attributes can be divided into two categories: static and dynamic. Static attributes are those that typically do not change during the execution of the project. These attributes are derived from the project specifications. Examples of static attributes include project goal, cost, duration, and number of tasks.

Dynamic attributes are those that change during the execution of the project. As such, they define the current state of the project. Examples of dynamic attributes include resources utilized, time elapsed, and number of tasks completed.

A project is completely defined by the project's specification. However, the static attributes defined in the specification do not provide the contractor or the owner with information needed for monitoring and controlling the progress of the project. For this reason, dynamic attributes are needed in project management.

4.2.5   States of a Project

All projects have two well-defined states that are documented in the project specification: the state at the start of the project and the state desired when the project is completed. Between these two states there are several intermediate states that are often difficult to define. These intermediate states, however, must be determined objectively and consistently to monitor the progress of a project. If a current state indicates that the project is not on track then corrective measures may be taken to rectify the situation. Thus, ascertaining the current state of a project is essential for effective project management.

A project's state can be determined from any one of or a combination of the project's dynamic attributes. Figure 4.4 shows a typical plot of resource utilization versus time elapsed for a given project. Comparing the as-execution project progress plot with the as-planned progress plot will provide the project manager with useful information about the progress of the project. The time elapsed need not be the abscissa of these plots. Figure 4.5 shows a plot of resource utilization versus the number of tasks completed. All two-dimensional project state plots are monotonically increasing graphs.

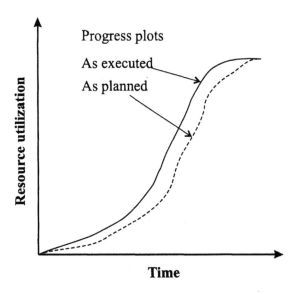

Figure 4.4   Typical resource utilization versus time state diagram for a project

They may be continuous or discrete depending on the attribute used. In any case, the rate of change of an attribute indicated in these plots and their deviation from the as-planned plots provides an objective measure of the current state of the project and its relationship with the as-planned state.

4.2.6   Project Time and Cost

Time and cost are the two most important attributes of a project. However, these attributes are not independent of each other. As a general rule, a faster completion of a project usually requires more resources. The static project duration and cost are determined as a trade-off by the parties before the start of the construction phase. Once these have been determined they serve as target values for the

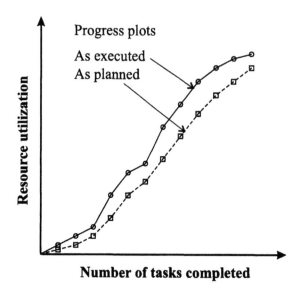

Figure 4.5 Typical resource utilization versus number of tasks completed for a project

dynamic project attributes. A project manager can then focus on achieving these targets by monitoring the progress of the resource utilization versus time plot. This plot provides essential information to the project manager regarding the current state of the project and its future trend, which is invaluable for project control.

Because of the importance of the project time and cost, it is essential that these values are determined as accurately as possible during planning. Therefore, an essential component of project planning is time and cost estimation. A plan is as accurate as the estimates of the time and cost.

4.3 PROJECT PLANNING AND MANAGEMENT

4.3.1 Introduction

The process of guiding a project from its conception to its completion to the satisfaction of all the parties involved is called project planning and management. This is a collective responsibility that falls on all the participants. However, the owner has the greatest responsibility and authority in the successful outcome of the project. Project planning and management requires both information creation and information sharing. As such, this process lends itself to concurrent and collaborative engineering (Figure 4.1). Its solution therefore requires sound engineering principles for information analysis and decision making. Further, the process should be automated to facilitate reliability and reproducibility of the results, which is essential in a collaborative environment.

In this book, project planning and management is considered to be the planning and management of the construction phase of the project's life cycle. As stated earlier, the construction phase is the most challenging phase of a project contributing to the overall success of the project directly and significantly. The models presented in this book, however, can be applied to other phases of the project's life cycle as well. Further, in the developments of the models we focus on the owner's view of the project. The owner's view is broader in scope and encompasses the views of most project participants, including the contractor.

4.3.2   Component Models

Project planning and management is made up of several components, each representing an essential process or procedure in the overall planning and management process. The components and the relationships among them are shown schematically in Figure 4.6. The integration and robustness of these component models collectively determine the effectiveness of the overall project planning and management. For example, a sophisticated scheduling model would be rendered ineffective if the time estimates used are unreliable. Similarly, no matter how sophisticated the time and cost management procedures for a project are, they would be ineffective if the scheduling model cannot capture the project in sufficient detail free of unrealistic simplifications. As another example, the change order management model depends directly or indirectly on all other component models. This relationship and inter-dependence of component models highlight the need for all the models to be reliable and robust. To summarize, effective project planning and management requires the following two conditions to be satisfied: (1) robustness and accuracy of the component models and (2) integration and interoperability of the component models.

The component models shown in Figure 4.6 are described briefly in the following paragraphs.

Time and Cost Estimation

Time and cost estimates are the building blocks of project planning and management. Estimates are used to ascertain project feasibility

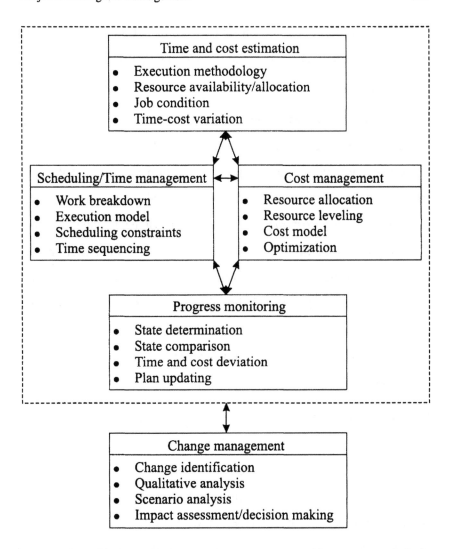

Figure 4.6   Project planning and management components and their relationships

and to develop and maintain detailed schedules and plans. In current industry practice, time and cost estimates are determined somewhat arbitrarily by schedulers based on their understanding of the conditions. In this book, we present a new model for cost estimation based on historical data. This is described in Chapter 10.

Scheduling

The process of creating and maintaining a plan of work that documents the sequence and timetable of execution is called scheduling. As such, scheduling considers only the time attribute of the project and not the cost. However, as explained in the previous section, time and cost are not independent of each other. Current commercial scheduling software systems ignore the construction cost during scheduling. This can lead to poor planning and management. Further, the scheduling model should be general and flexible in order to capture the actual execution conditions on site. In this book, we present a new integrated scheduling and cost management model for project management, starting with Chapter 5.

Cost Management

Project cost management ensures that the project is completed within budget. As such, issues such as work crew and equipment sequencing, allocation, and distribution have to be considered. These issues should be considered together with the time and schedule of the project as both are of utmost importance to the project participants. Further, the project participants desire a plan that has the minimum cost. For these reasons we incorporate cost optimization in our scheduling model to produce reliable minimum cost schedules. The use of the neural dynamics optimization algorithm for construction minimization is described in Chapter 6.

Progress Monitoring

Progress monitoring is an integral part of any project control and management problem. In project management, progress monitoring essentially involves two tasks: (1) determining the current state of the project and (2) determining the deviation of the current state from the as-planned state of the project at a given time. This information can then be used to plan and control the future progress in such a way that the project goals are satisfied. Progress monitoring requires accurate and consistently reliable state determinations. This in turn requires schedules that accurately represent the state of the project at any given time. In this book, progress monitoring is based on the scheduling model developed in Chapters 5 and 6 and the software implementations presented in Chapters 8 and 9.

Change Order Management

Change management ensures that change orders initiated in a project during the construction phase are handled fairly and reliably. This requires objective decision making capabilities derived from reliable information about the state of the project. The handling of change orders is described in Sections 7.1–7.3, 8.2.3, and 9.6.

4.4   ELEMENTS OF PROJECT SCHEDULING

In this section basic elements of project scheduling are introduced and described briefly. Essentially, there are four major steps in the preparation of a schedule:

8. Creating the work breakdown structure where the total work needed for the project is divided into recognizable and logical chunks of work called tasks.

9. Assigning resources to the tasks and estimating their durations.

10. Specifying scheduling constraints, which can be resource constraints or logic constraints between the tasks.

11. Generating the optimal schedule by using an algorithm to appropriately sequence and time each task in the project.

4.4.1   Tasks

A task is a well-defined job or activity that consumes resources and requires time for its completion. A task is distinguished from other tasks by its resource requirements (labor and equipment, collectively known as crew), material requirements, and goal. Each task in a project is given a unique identification and a title that briefly describes its goal or work. In many construction industries, common recurring tasks that have well-defined goals and resource and material requirements are predefined and documented in manuals of construction practice. For example, compaction where self-propelled pneumatic rollers are used will be considered a different task than the one where manual compactors are used. In this example, a different set of labor and equipment mandates that a different task be defined so that each can be scheduled independently, even though the two tasks have the same goal. Work that can be scheduled independently must be defined as separate tasks to ensure that the resulting schedules are the most efficient.

A task may be non-repetitive or repetitive. A task is repetitive if the same job or activity is performed at different times and/or at different locations. Each part of the repetitive task, called a segment of work, utilizes the same resources and has the same qualitative goal. Modeling a task as repetitive rather than a set of non-repetitive tasks has several advantages such as control over continuity of work, use of multiple-crew strategies, distance and time modeling of work, and reduction in schedule complexity. Examples of repetitive tasks include construction of floors in a highrise building project and resurfacing operation in a highway rehabilitation project.

A task i, or a segment of work in a repetitive task identified by i, is completely defined for scheduling purposes by the start time of the task denoted by T_i and the duration of task denoted by D_i. The finish time is then given by $T_i + D_i$. The durations are input values estimated from a consideration of the resources allocated to the task and the job conditions, whereas the start times are determined by the scheduling algorithm.

4.4.2 Work Breakdown Structure

The first task in the preparation of a schedule is the breakdown of the project's work into smaller chunks. Each of these chunks of work represents a fraction of the entire project's work. Collectively the chunks of work must capture all the project's work without any omissions. This completeness of work is the first basic requirement for an accurate work breakdown. The second basic requirement is that the divisions of work be well defined or, in other words, be classified as tasks. Thus, the division of work cannot be arbitrary but has to follow the definition of tasks so that practically meaningful divisions are produced.

The division of a project's work into tasks is not well defined and different schedulers may come up with different sets of tasks. If a manual of construction practice is available for that particular type of project then the scheduler can use the predefined tasks in the manual if they match with any of the project's work. Effective work breakdown into tasks requires good experience of construction methodology and practice. The general procedure for doing this is to start with the project's goal and successively sub-divide it into sub-goals until a set of goals are obtained that can be defined as tasks. When represented graphically this procedure produces a hierarchical structure called a work breakdown structure (Figure 4.7) (Naylor, 1995). Each node represents a goal whereas its child nodes represent sub-goals that are collectively equivalent to the parent's goal. The top node in the hierarchy represents the overall project goal and the leaf nodes represent the goals or tasks into which the project has been divided for scheduling purposes.

When should a scheduler stop further subdivisions of work? There are two general rules for this, both requiring an understanding of the construction process. First, the leaf nodes must represent a task; that is, they constitute a chunk of work that is commonly considered in construction practice as a unit with a specific goal. Second, the subdivision should be carried to such an extent that no

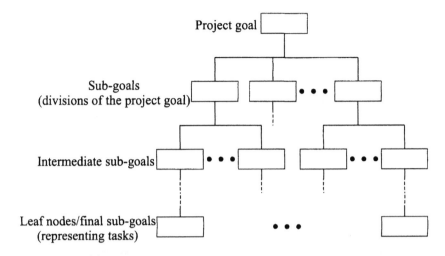

Figure 4.7   Hierarchical   layout   of   a   typical   work   breakdown   structure

important scheduling constraint is lost. That is, a task should be subdivided if an important constraint applies to only a portion of the task's work. If no such constraint exists then there is no need to further subdivide, as this will only increase the complexity of the schedule without increasing accuracy. Figure 4.8a shows a simple work breakdown structure for the construction of a building. Only two tasks are identified: construction of the substructure and construction of the superstructure. This, however, is not an accurate work breakdown as several important constraints that influence work within each task are lost. Figure 4.8b shows a more detailed work breakdown structure where the substructure construction task has been subdivided further. This breakdown is more effective because, for example, the constraint that excavation should occur before laying the foundation can be modeled.

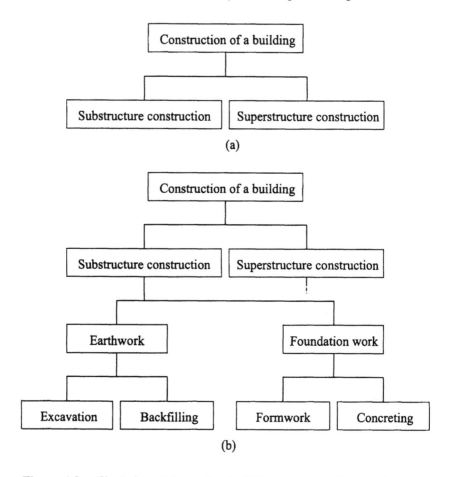

Figure 4.8   Work breakdown for building construction project: (a) An inadequate breakdown, and (b) An adequate subdivision of substructure construction

For repetitive tasks the work is further subdivided into work crews and segments of work assigned to each crew. The breakdown of repetitive tasks will be described further in Chapter 5.

4.4.3   Scheduling Constraints

The tasks constituting a project cannot be executed in any arbitrary order but are constrained by "natural" work ordering, construction methodology, practical feasibility, and resource availability. These constraints are either specified between two tasks or between one or more tasks and an absolute value. Scheduling constraints should be specified with great care. Specifying an unnecessary constraint that does not exist in reality or omitting an important constraint in the model can completely invalidate the scheduling results. An understanding of the construction procedures in practice is therefore essential.

Scheduling constraints can be grouped under three headings, described in the following paragraphs.

<u>Logic Constraints</u>

Logic constraints are specified between two tasks so that one task precedes or follows another task in time. These constraints, also known as precedence relationships, are based on either the start or the finish time of the two tasks. As such, four types of relationships are possible: finish-to-start (FS), start-to-start (SS), finish-to-finish (FF), and start-to-finish (SF). Mathematically, these constraints are expressed by inequality constraints as follows:

Finish-to-start (FS)

$$T_i + D_i + L_{FSij} \leq T_j$$

Start-to-start (SS)

$$T_i + L_{SSij} \le T_j$$

Finish-to-finish (FF)

$$T_i + D_i + L_{FFij} \le T_j + D_j$$

Start-to-finish (SF)

$$T_i + L_{SFij} \le T_j + D_j$$

In these equations, the indices i and j refer to the preceding and following tasks, respectively, and L is the time lags/leads or slack times allowed in the constraint. In words, the finish-to-start relationship, for example, means that task j can only start after L time units before (if L is negative, i.e., it is a lag) or after (if L is positive, i.e., it is a lead) the completion of task i. As an example, a finish-to-start relationship exists between the task "site clearing" and the task "site office construction." Even when no explicit logic constraint is specified between two tasks one may exist by virtue of constraints specified between other tasks. For example, if an FS relationship exists between tasks A and B and between tasks B and C, then an FS constraint is implied between tasks A and C. The scheduler has to be careful while specifying logic constraints so that no non-existent constraints are introduced.

Absolute Constraints

An absolute constraint specifies a time constraint on the finish and/or start time of a task. An absolute constraint is commonly specified on all the tasks to ensure that they end before the project completion date/time. Mathematically, this can be written as

$$T_i + D_i \leq D^{\max} \quad \text{for all } i$$

where $D^{\max}$ is the maximum project duration. Absolute constraints can also be specified at intermediate stages of the project when completion of some tasks is desired at certain times. These times represent milestone or time points in the project's progress.

Buffer Constraints

Buffer constraints are specified between two tasks to ensure that a minimum buffer is maintained between them throughout their execution. The buffer may be a time buffer or a distance buffer. For example, a time buffer is needed between the laying of a pipeline and back filling the trench to allow time for concrete anchors to cure. Buffer constraints are most useful in modeling repetitive tasks. They will be described further in Chapter 5 where the new scheduling model is presented.

4.5   GRAPHICAL DISPLAY OF SCHEDULES

4.5.1   The Need

Project schedules contain essential information for project planning and management. This information, however, is not useful unless it is understood and applied correctly by the project manager. Presenting the information in a manner that is easily and unambiguously understood is therefore very important. A typical

construction project schedule may have hundreds or even thousands
of tasks each having several attributes. Furthermore, the tasks in a
schedule are often related by complex scheduling logic. A non-
graphical or text output of the schedule will overwhelm the project
manager. Also, a text representation cannot effectively convey to the
project manager the overall state of the project, the scheduling
constraints, and the task precedence relationships. In other words,
graphical displays are necessary, if not as the primary medium of
schedule communication, then certainly as a supplement to the text
representation.

Several graphical representations of a project are possible. Each
has advantages and disadvantages, and some are more widely
recognized and accepted in the industry than others. In the following
sections, we describe three popular graphical representation formats
commonly used in practice.

4.5.2   Gantt Charts

Gantt charts are simple bar charts drawn on a time scale. Each bar
represents a task in which the left and the right edges of the bar
indicate the start and the finish time of the task. The Gantt chart does
not show the precedence relationships between the tasks. As such, it
is not very useful for significant construction projects where complex
relationships often exist between tasks. The usefulness of Gantt
charts is therefore limited to small projects or to a small portion of a
large project. However, Gantt charts are easy to understand and are

good for presentation purposes. Gantt charts are not recommended as the primary tools for monitoring and managing projects.

4.5.3   Network Diagrams

The network diagram is the most popular medium for the graphical presentation of project schedules (Moder and Phillips, 1970; Willis, 1986; Hendrickson and Au, 1989; Naylor, 1995). Network diagrams consist of nodes, arrows, and lettering and capture both task times and logic constraints. There are two different ways of representing schedules as a network of nodes and arrows. In an activity-on-arrow (AOA) diagram, arrows are used to represent tasks whereas nodes represent events or times of importance such as the start and finish time of tasks. In an activity-on-node (AON) diagram, on the other hand, nodes represent tasks and arrows between tasks establish precedence relationships. The AOA and AON diagrams are equivalent in their modeling capabilities; a schedule that can be represented by an AOA diagram can also be represented by an AON diagram, and vice versa. The AON diagram, however, is more general, compact, and displays more information than the AOA diagram. We use AON diagrams in this book and refer to them as either network diagrams or activity-on-node diagrams.

Each node in an activity-on-node diagram represents a task. The general layout for displaying a task in an AON diagram is shown in Figure 4.9. The task identification, description, and time attributes are written inside a rectangular shaped box. The task identification is required information and the rest is optional. However, the display of

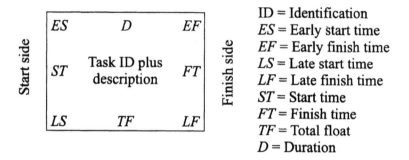

Figure 4.9   Representation of a task (node) in an activity-on-node diagram

the task description, duration, start time, and finish time is strongly recommended for the diagram to be effective. With the exception of the task duration all other time attributes are computed using the scheduling algorithm.

A logic constraint or relationship between two tasks is represented by a directed arrow from one node to the other. The node to which the arrow points is the constrained task whereas the node from which it emerges is the constraining task. The type of the constraint is determined by the ends of the nodes to and from which the arrow points. For example, if the arrow emerges from the finish side of node i to the start side of node j, then a finish-to-start (FS) relationship exists between tasks i and j. The identification written above the arrow uniquely identifies the constraint, and if the constraint has a lag time this information is written under the arrow. Figure 4.10 shows AON diagrams for the four logic constraints defined in the previous section. Logic constraints that are governed by parameters other than lag time such as quantity of work can also

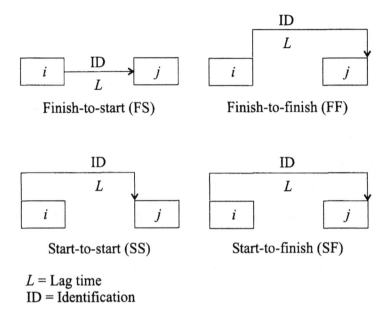

Finish-to-start (FS)   Finish-to-finish (FF)

Start-to-start (SS)   Start-to-finish (SF)

L = Lag time
ID = Identification

Figure 4.10 Representation of logic constraints in activity-on-node diagrams

be represented by clearly indicating the rule under the arrow. The AON diagram cannot represent buffer or resource constraints. It also cannot represent repetitive tasks. Note that an arrow represents an explicit constraint between two tasks; implicit constraints may exist between tasks as a result of the explicitly specified constraints.

The AON diagram is widely used in practice and is considered a de-facto standard in the construction industry. Most scheduling software programs are capable of generating network diagrams for the display of schedules. Graphical enhancements like color, varying thickness of lines, and time-scaled drawing may be used to convey more information like the critical path in a network.

4.5.4   Linear Planning Chart

The linear planning chart (LPC) displays tasks on a two-dimensional scaled plot of time on one axis and an appropriate space attribute (e.g. distance) on the other. As such, the LPC captures both the temporal and the spatial attributes of the tasks in a project. Tasks in an LPC are represented by line segments drawn to scale on the plot indicating their start and finish times, and their start and finish locations (Figure 4.11). Each task is labeled by its unique identifier. LPCs are a valuable representation approach for linear projects such as highway, tunnel, and pipeline construction where the location of work plays a significant role in planning and management. With the help of the LPC, the project manager of a linear project can determine the current state of work, the rate of progress of work, and the spatial locations of tasks. The rate of progress of a task at any given time is given by the slope of the line segment representing that task at that time.

Time and distance buffer constraints can readily be displayed on an LPC. Figure 4.12 shows a time buffer constraint between tasks i and j and a distance buffer constraint between tasks j and k. A time buffer constraint ensures that a minimum time lag exists between two tasks at a given location. Similarly, a distance buffer constraint ensures that a minimum distance buffer is maintained between the work of the two tasks at all times. Using directed arrows, logic constraints can also be displayed on a LPC, just as in a network diagram. However, the display may become cluttered and difficult to

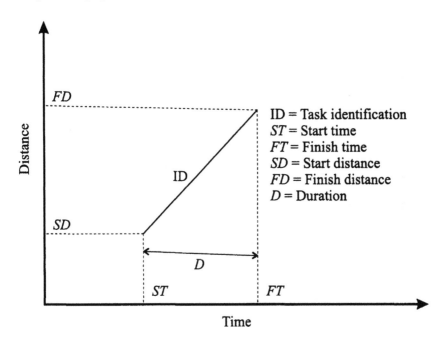

Figure 4.11 Representation of a task on a linear planning chart

read and understand. Legibility can be improved by using graphical symbols, color, and varying the thicknesses of lines.

Linear planning charts are not widely used in practice. This is primarily because the critical path method—the most widely used scheduling technique today—cannot generate the information needed for an LPC. Unlike the network diagram, the LPC can represent repetitive tasks, buffer constraints, multiple crews per task, work continuity constraints, location models, and progress rates of tasks. And as mentioned earlier, the LPC is an excellent presentation tool for linear projects, providing an intuitive understanding of the project's work. The LPC is therefore most appropriate for displaying

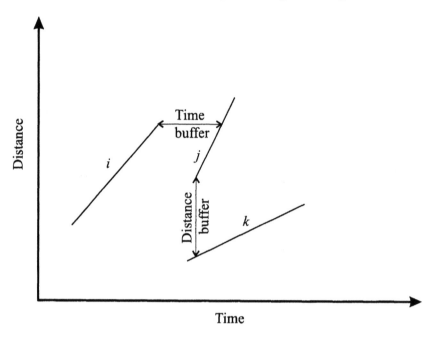

Figure 4.12 Representation of buffer constraints on a linear planing chart

schedules generated by advanced scheduling techniques that can take advantage of its information communication features. Because logic constraints are sometimes difficult to display on a LPC, network diagrams may be used concurrently with LPCs.

4.6   THE CRITICAL PATH METHOD

4.6.1   Introduction

The critical path method (CPM) is a scheduling technique based on a network diagram representation of the project. Developed in the late 1950s for time sequencing manufacturing operations, the CPM quickly became popular in other application areas including construction project scheduling (Moder and Phillips, 1970). Nowadays, the CPM is considered the de-facto standard in the construction industry and is implemented in several widely-used commercial software systems for construction project planning and management. The CPM takes as input the durations of tasks and the logic constraints (precedence relationships) between tasks, and produces as output a range of start and finish times for each task that results in the minimum project duration satisfying the logic constraints. The CPM also identifies a critical path—a continuous path or paths through the network of nodes and arrows whose length is equal to the project duration. Any delay along this path will cause a delay in the overall project duration. The CPM consists of simple arithmetic operations only.

The original CPM was only capable of handling finish-to-start precedence relationships (Moder and Phillips, 1970). Later developments added the capability to handle all four logic constraints (Willis, 1986; Hendrickson and Au, 1989). Sometimes the name precedence method is used for the enhanced CPM (Willis, 1986). In this book, CPM refers to the enhanced algorithm, which is generally the case in practice. The project evaluation and review technique (PERT) is another scheduling technique also developed in the late 1950's. PERT is fundamentally similar to the CPM, being based on a network diagram representation of the project. Unlike the CPM, however, the PERT uses as inputs a statistical distribution for the task durations and the time outputs have statistical significance. PERT is not widely used in the construction industry.

4.6.2   Features

CPM is a simple scheduling algorithm that is based on a simple project model. For the CPM algorithm to be applicable, a network diagram (the activity-on-node diagram) must be able to represent the project. This condition defines the extent of the schedule model that CPM can handle. As such, CPM suffers from the same shortcomings noted for the AON diagram in Section 4.5.3. The CPM scheduling model is briefly outlined below.

- The project is made up of non-repetitive tasks only. Each task is a single entity of work with no subdivisions.

- A task i is completely defined by its duration, D_i, which is a pre-assigned fixed value.

- All four logic constraints (FS, SS, SF, FF) can be specified between tasks in a project.

- Location modeling of tasks, multiple crew strategies, buffer constraints, time variation with job conditions, and cost of tasks are not handled explicitly.

Once the project schedule-model is developed the CPM algorithm can be used to calculate the task times that satisfy the constraints. A CPM analysis produces the following outputs:

- The early start (ES_i), early finish (EF_i), late start (LS_i), and late finish (LF_i) times for each task i.

- The minimum project completion time or duration.

- An identification of the tasks and the relationships that are critical and must be completed on time to avoid project delay.

- The float or available time flexibility in the start or finish time of a task that will not delay the project's completion.

The CPM requires only addition and subtraction operations for computations.

4.6.3   Parameter in the CPM Analysis

Terms and parameters commonly used in CPM analysis are defined in this section (Willis, 1986; Hendrickson and Au, 1989).

<u>Task Times</u>

As mentioned earlier, the CPM computes four time values for each task. The early start time (ES_i) is the earliest time at which a task can be started given the durations and constraints specified. The early finish time (EF_i) is equal to the early start time plus duration $(ES_i + D_i)$. The late start time (LS_i) is the latest time a task can be started without delaying the project completion. The late finish time (LF_i) is equal to the late start time plus duration $(LS_i + D_i)$.

<u>Task Floats</u>

Task floats are allowable time variations within which a task can start or finish without delaying the project. There are three types of task floats that determine the criticality of the start time, the finish time, or both to the timely completion of the project. The total float for task i (TF_i) is defined as

$$TF_i = LF_i - ES_i - D_i$$

A task is critical if its total float is zero. That is, the task must be started at its early start time, executed for its duration, and be completed by its late finish time.

The starting float for task i (StF_i) is defined as

$$StF_i = LS_i - ES_i$$

The start of a task is critical if its starting float is zero. That is, there is no flexibility available in the starting of the task, and any delay will produce a delay in the project. If a task is critical then its starting is also critical. However, the inverse is not always true.

The finishing float for task i (FnF_i) is defined as

$$FnF_i = LF_i - EF_i$$

The finish of a task is critical if its finishing float is zero. In other words, the task must finish by its late/early finish time; otherwise, the project will be delayed. If a task is critical then its completion is also critical. However, the inverse is not always true.

Relationship Float

A logic constraint between two tasks, represented by an arrow in an activity-on-node diagram, can also be critical. This will be the case when the two sides of the task that are part of the relationship are also critical. The criticality of a relationship m is given by its relationship float (RF_m) defined as follows:

$$RF_m = LS_j - EF_i - L_m \quad \text{for FS relationship}$$

$$RF_m = LS_j - ES_i - L_m \quad \text{for SS relationship}$$

$$RF_m = LF_j - ES_i - L_m \quad \text{for SF relationship}$$

$$RF_m = LF_j - EF_i - L_m \quad \text{for FF relationship}$$

where the indices i and j represent the tasks at the tail and head ends of the arrow, respectively, and L_m is the lag time for logic constraint m.

Critical Path

The critical path is the continuous connected path through the network made up of critical tasks, critical starting of tasks, critical finishing of tasks, and critical relationships. The critical path may be a single path or may branch out into two or more paths. The duration of the critical path gives the duration of the project. A delay at any point along the critical path will delay the project's completion.

4.6.4 Algorithm

In construction textbooks (Moder and Phillips, 1970; Willis, 1986) CPM analysis is usually described graphically by a sequence of steps and operations performed on the project's network diagram. This approach, however, is not rigorous for computer implementation, which nowadays is the primary tool used for CPM analyses. For this purpose, we present a detailed step-by-step algorithm for the CPM in this section.

Let N be the total number of nodes (tasks) in the network and R_i be the total number of relationship arrow heads or tails (constraints) that originate or terminate, respectively, on node i. The CPM consists of five phases.

Phase 1: Node Numbering

Number all N nodes consecutively such that the number assigned to a node at the head of an arrow is always greater than that at the tail of the arrow.

Phases 2: Forward Pass

Compute the early start and the early finish times for tasks.

1. For each node j ($j = 1$ to N) do steps 2 to 6.

2. Set $ES_j^k = EF_j^k = 0$ where $k = 1, R_j$.

3. For each arrow k terminating on node j ($k = 1, R_j$) do steps 4 and 5.

4. Calculate the early time at the head of the arrow. The early time is the early start time when the relationship is FS or SS and the early finish time when the relationship is FF or SF. The appropriate early time is found by algebraically adding the appropriate early time at the tail of the arrow to the relationship lag. For example, for an FS relationship, compute $ES_j^k = EF_i + L_k$ where i is the node at the tail of relationship k.

5. Calculate $EF_j^k = ES_j^k + D_j$.

6. Calculate the final early start and early finish times as the maximum of all values computed in steps 4 and 5. $ES_j = \max_{k \in R_j}\{ES_j^k\}$ and $EF_j = \max_{k \in R_j}\{EF_j^k\}$.

Phase 3: Project Duration

Compute the project duration, D^{proj}, as equal to the maximum early finish time value computed in phase 2: $D^{proj} = \max_{i \in N}\{EF_i\}$.

Phase 4: Backward Pass

Compute the late start and the late finish times for tasks.

1. For each node j ($j = N$ to 1) do steps 2 to 6.

2. Set $LS_j^k = LF_j^k = 0$ where $k = 1, R_j$.

3. For each arrow k originating from node j ($k = 1, R_j$) do steps 4 and 5.

4. Calculate the late time at the tail of the arrow. The late time is the late start time when the relationship is SS or SF and the late finish time when the relationship is FS or FF. The appropriate late time is found by algebraically subtracting the relationship lag from the appropriate late time at the head of the arrow. For example, for an FS relationship, compute $LF_j^k = LS_i - L_k$ where i is the node at the head of relationship k.

5. Calculate $LS_j^k = LF_j^k - D_j$.

6. Calculate the final late start and late finish times as the minimum of all values calculated in steps 4 and 5.

$$LS_j = \max_{k \in R_j}\left\{LS_j^k\right\} \text{ and } LF_j = \max_{k \in R_j}\left\{LF_j^k\right\}.$$

Phase 5: Critical Path Identification

Using the equations presented in Section 4.6.3 the task and relationship float values are calculated from the computed task times. Critical tasks, critical starting time of tasks, critical finishing time of tasks, and critical relationships are usually identified and marked as thick lines on the network diagram. The critical path consists of all critical elements that form a connected path through the network. The duration of this path is equal to the project duration.

4.6.5   Example

The construction of a single-story steel-framed office building is used to illustrate the CPM. The work for the project is divided into 15 tasks linked by 20 logic relationships. The description of tasks and their durations are given in Table 4.1 while the logic relationships are defined in Table 4.2 The activity-on-node diagram for the project is shown in Figure 4.13. To model concurrent operation of two tasks a combination of start-to-start and finish-to-finish relationships is used between them. For example, an SS and FF relationship with lag times of 1 day is used between task 3 (Place formwork and reinforcement) and task 4 (Pour concrete) to simulate parallel execution. The time lag of 1 day ensures that formwork and reinforcement are in place before concreting takes place. Finish-to-start relationships are used to model non-overlapping operations. For example, an FS relationship is specified between task 6 (Backfill and

Table 4.1   Description and duration of tasks for the steel building project

Task #	Description	Duration (days)
1	Clear site	3
2	Excavate for footings	6
3	Place formwork and reinforcement	3
4	Pour concrete	3
5	Remove formwork	1
6	Backfill and compact	5
7	Fabricate steel frame	10
8	Secure steel frame	8
9	Secure prefabricated roofing	5
10	Lay flooring	5
11	Erect masonry fill-in	8
12	Attach doors and windows	3
13	Electrical work	5
14	Plumbing	5
15	Finish interior	12

compact) and task 10 (Lay flooring) to simulate the natural ordering of these tasks.

The results of the CPM analysis are shown in Figure 4.13. The displayed information includes the task times, task total floats, the project critical path, and the project minimum duration. The minimum project duration found by CPM analysis is 66 days. As can be seen from the network diagram, most tasks in this project are critical and therefore must be completed on schedule to avoid a delay in the project's completion. Task 7 (Fabricate steel frame), for example, is not critical and has a total float of 18 days. This means that the execution of this task can be delayed by up to 18 days, and provided that the task ends before the 28th day the project will not be

Table 4.2   Definition of logic relationships in the steel building project

Relationship #	Type	Preceding task #	Following task #	Lag (days)
1	FS	1	2	0
2	SS	1	7	0
3	FS	2	3	0
4	SS	3	4	1
5	FF	3	4	1
6	SS	4	5	3
7	FF	4	5	3
8	FS	5	6	7
9	FS	6	8	0
10	FS	7	8	0
11	FS	8	9	0
12	FS	9	10	0
13	SS	10	14	2
14	FF	10	14	2
15	FS	10	11	0
16	SS	11	13	0
17	FF	11	13	0
18	FS	11	12	0
19	FS	14	15	0
20	FS	13	15	0

delayed. The completion of tasks 5 and 13 is critical meaning that these tasks must be completed at their early/late finish times to avoid a project delay. However, the start time of these tasks is not critical.

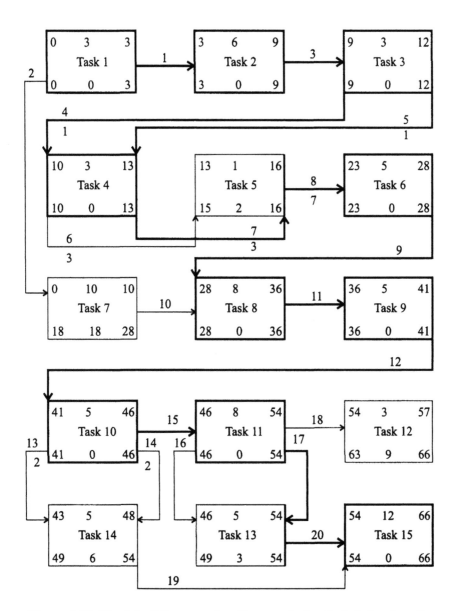

Figure 4.13 Steel building construction project scheduled using the CPM

5

A GENERAL MATHEMATICAL FORMULATION
FOR PROJECT SCHEDULING AND COST
OPTIMIZATION

5.1   INTRODUCTION

Most construction projects involve a combination of repetitive and non-repetitive tasks. A typical example is highway construction in which tasks such as clearing and grubbing are performed repeatedly over the length of the highway, and tasks such as site office construction are carried out only once. Presently, traditional network scheduling methods such as CPM and PERT are used for the scheduling and monitoring of such projects. Despite their extensive use these methods have a number of shortcomings:

- Network methods do not guarantee continuity of work in time which may result in crews being idle.

- Multiple-crew strategies are difficult to implement in the network methods.

139

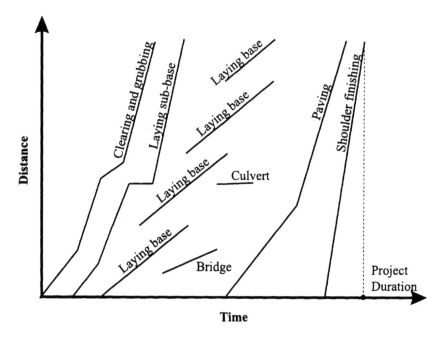

Figure 5.1   Linear planning chart

- The network diagram is not suitable for monitoring the progress of a project.

- Network methods do not provide an efficient structure for the representation of repetitive tasks. All tasks are represented similarly and there is no consideration of the location of work in the scheduling.

To overcome these shortcomings new approaches have been proposed in the literature particularly for repetitive projects. Figure 5.1 presents a linear planning chart, which is a graph of location (distance) versus time for the work to be carried out. Such a planning chart represents the progress of a task and can be used to monitor a project. The linear planning chart (also called LSM diagram) motivated the development of linear scheduling method (LSM). Selinger (1980) presented equations for the lines in a linear planning chart assuming non-interference of crews and continuity of work. Johnston (1981) presented applications of the LSM for highway construction projects. Using optimal control theory, Handa and Barcia (1986) formulated the problem as an optimization one minimizing the project duration. These early LSM models had limitations such as constant rate of production for each task, binding continuity constraints, and no provisions for the use of multiple crews.

Russell and Caselton (1988) presented a dynamic programming formulation to minimize the project duration. Their formulation can accommodate variable production rates for each task and non-binding work continuity constraints. Russell and Wong (1993) described and showed the use of a general scheduling model developed by incorporating the capabilities of CPM and LSM. In their model, each task is defined by a set of attributes which are then linked together using general precedence conditions to form a schedule.

Civil engineering construction projects are large projects in terms of capital requirement. Minimizing cost is therefore a primary goal in the planning and scheduling of such projects. Cost, however, is closely related to time. In general, direct project cost increases with a decrease in project duration and this trade-off problem is complicated by the number of variables involved. A computer model to automate the process of project direct cost minimization is therefore highly desirable.

In the more recent literature, attempts have been made to model direct cost optimization of construction projects using LSM and CPM. These systems, however, have limited practical use because they are based on simple scheduling models. Reda's (1990) LSM model assumes constant production rate for each task and binding constraints on work continuity. The cost-duration relationship for each task is assumed to be linear. As such, the model is useful only for simple and straightforward repetitive projects where all tasks have the same production rate which is rare in practice. Further, a project having both repetitive and non-repetitive tasks cannot be handled. Liu *et al.* (1995) present a hybrid linear/integer programming approach for handling a combination of discrete and linearly continuous cost-duration relationships for tasks. It is based on the CPM scheduling model and thus suffers from the same CPM's shortcomings in modeling of the construction project.

In this chapter, a general mathematical formulation is presented for scheduling of construction projects. Various scheduling constraints are expressed mathematically. The construction scheduling is posed as an optimization problem where project direct cost is minimized for a given project duration assuming any combination of linear and nonlinear task cost-duration relationships.

In the next chapter, the robust neural dynamics model of Adeli and Park (1995b) is adapted for optimization and solution of the problem.

5.2 COST-DURATION RELATIONSHIP OF A PROJECT

The major cost of a project consists of direct and indirect costs. The resources allocated to each task of a project determine the direct cost. Indirect costs are overhead costs. Additional costs may be incurred by the contractor in the form of damages if the project is not completed on time. The duration of a project is obtained by sequencing individual tasks whose durations are estimated from a knowledge of the resources allocated to each task and the job conditions. Thus, cost and duration are intricately related. Both of these parameters are of great importance to the contractor who strives to minimize the cost while at the same time satisfying the contractual requirements, the most important of which is the completion deadline. Assuming the sequencing constraints are not changed, the direct cost, in general, has an inverse relationship with the duration of a construction project. The indirect cost increases with an increase in the duration of the project. The total cost is the sum of these two and can increase or decrease with duration. The goal is to obtain the global optimum solution for the scheduling problem.

There is also an inverse relationship between direct cost and the duration of an individual task. A scheduler estimates the time required to complete a task from the resources allocated to it. This

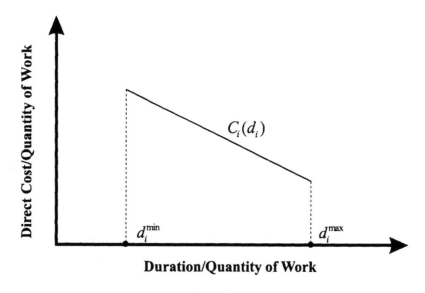

Figure 5.2   Linear direct cost-duration curve for unit quantity of task *i*

time is based on assumed labor and equipment productivity rates ignoring the effects of varying job conditions. Depending on the options and the availability of resources the scheduler has for each task, a cost-duration curve can be constructed. This curve can be continuous or discrete. For efficient mathematical formulation the discrete relationship is approximated by a continuous linear (Figure 5.2) or nonlinear (Figure 5.3) curve. In this way a continuous variable optimization technique can be used to solve the construction time-cost trade-off problem.

In projects with repetitive tasks such as highway construction, it is convenient to represent cost and duration of a task in unit quantities of work. If d_i is the time required to complete a unit quantity of work of task i and W_{ij} is the total quantity of work

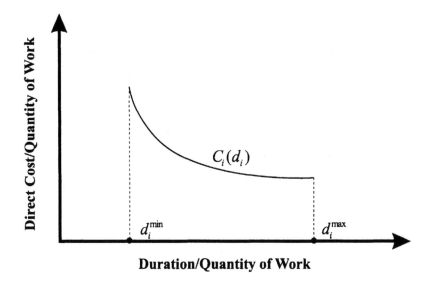

Figure 5.3   Nonlinear direct cost-duration curve for unit quantity of task *i*

required in segment j of task i, then the actual duration, D_{ij}, can be expressed as:

$$D_{ij} = \mu_{ij} d_i W_{ij} \tag{5.1}$$

where μ_{ij} is the job condition factor reflecting the effects of variable conditions such as weather, soil conditions, terrain, site congestion, learning effects, etc.

5.3   FORMULATION OF THE SCHEDULING OPTIMIZATION PROBLEM

A general mathematical formulation of the construction scheduling problem is presented in this section. The advantage of such a general formulation is that it can be specialized and reduced for the solution of specific and perhaps less complicated scheduling problems. Further, it can be effectively integrated with the general neural dynamics model for solution of optimization problems developed by Adeli and Park (1995b).

Both non-repetitive and repetitive tasks are considered in the formulation. The non-repetitive tasks correspond to the activities of the traditional network methods, such as CPM. A non-repetitive task involves no internal logic as it is performed only once. A repetitive task, on the other hand, may have an elaborate internal logic that connects the segments assigned to various crews. By specifying appropriate constraints, work continuity considerations and multiple-crew strategies can be modeled. A crew rarely performs at ideal productivity throughout; its performance is affected by the varying job conditions. This is included in our scheduling model by means of a factor, μ_{ij}, which modifies the ideal productivity of a crew to reflect the effect of the job conditions. The external logic of each task is specified by means of a full set of precedence relationships and/or stage (distance) and time buffers.

Development of the general scheduling formulation for a construction project such as highway construction involves the following steps divided into three main categories (headings) as follows:

5.3.1 Breakdown the Work into Tasks, Crews, and Segments

Step 1: Break down the project into N_T tasks. Identify non-repetitive and repetitive tasks. Let N_{NT} and N_{RT} be the number of non-repetitive and repetitive tasks, respectively. If $N_{RT} = 0$, skip steps 2 to 5 and go to step 6.

Step 2: For each repetitive task i, choose the number of crews to be used (N_{Ci}). Non-repetitive tasks have only one crew that performs over one segment only.

Step 3: Assign N_{Si}^k segments of the highway to crew k of repetitive task i. The segments are chosen considering the job conditions and quantity of work required, factors affecting the production rate. In addition, predetermined breaks in the work of a crew may influence the choice of segments. The segments are not required to have equal lengths or constructed in sequence. Each segment is identified by Z_{ij}^k

and Z^k , the beginning and ending distances at which repetitive task i is performed by crew k over segment j. Note that each crew of a task is assigned a unique set of segments; two crews cannot perform the same task over the same portion of the highway.

5.3.2 Specify the Internal Logic of Repetitive Tasks

For each crew of a repetitive task, do the following:

Step 4: Specify the work continuity relationship between segments j and $j+1$, in the following form:

$$T_{ij}^k + D_{ij}^k + S_{ij}^k \leq T_{i(j+1)}^k \tag{5.2}$$

where T_{ij}^k is the time at which crew k of task i starts work on segment j, D_{ij}^k is the duration of work for crew k of task i on segment j, and S_{ij}^k is the idle or slack time of crew k of task i between segments j and $j+1$. For continuity of work, S_{ij}^k must be equal to zero. If a task has only one crew skip step 5 and go to step 6.

Step 5: Define the start of a crew with respect to previous crew(s). The following precedence relationships of start-to-start, finish-to-finish, and start-to-finish are used:
Start-to-start (SS):

$$T_{i1}^k + L_{SSi}^{kl} \leq T_{i1}^l \tag{5.3}$$

Finish-to-finish (FF):

$$T_{iN_{Si}^k}^k + D_{iN_{Si}^k}^k + L_{FFi}^{kl} \leq T_{iN_{Si}^l}^l + D_{iN_{Si}^l}^l \tag{5.4}$$

Start-to-finish (SF):

$$T_{i1}^k + L_{SFi}^{kl} \leq T_{iN_{Si}^l}^l + D_{iN_{Si}^l}^l \tag{5.5}$$

where the superscripts l and k refer to the current and the previous crews, respectively; L_{SSi}^{kl}, L_{FFi}^{kl}, L_{SFi}^{kl} are the start-to-start, finish-to-finish, and start-to-finish time lags between crews k and l, respectively. These time lags may be given as a function of quantity of work and/or time. If more than one relationship is specified for a particular crew, only one will govern in the final minimum cost schedule obtained from the optimization algorithm. This particular relationship usually is not known in advance and all possible relationships have to be specified in the optimization model.

5.3.3   Specify the External Logic of Repetitive and Non-Repetitive Tasks

Step 6: Describe the sequencing of the tasks in the project. Each task can be linked with any number of previous tasks by specifying one or more of the following precedence relationships:

Start-to-start (SS):

$$T_{i1}^1 + L_{SSij} \leq T_{j1}^1 \tag{5.6}$$

Finish-to-start (FS):

$$T_{iN_{Si}^k}^k + D_{iN_{Si}^k}^k + L_{FSij} \leq T_{j1}^1 \quad k = 1,\dots,N_{Ci} \tag{5.7}$$

Start-to-finish (SF) (is specified when the task has only one crew):

$$T_i^1 + L_{SFij} \leq T_{jN_{Sj}^l}^l + D_{jN_{Sj}^l}^l \quad l = 1 \tag{5.8}$$

Finish-to-finish (FF) (is specified when both tasks have one crew only)

$$T_{iN_{Si}^k}^k + D_{iN_{Si}^k}^k + L_{FFij} \leq T_{jN_{Sj}^l}^l + D_{jN_{Sj}^l}^l \quad k = l = 1 \tag{5.9}$$

The quantities L_{SSij}, L_{FSij}, L_{SFij} and L_{FFij} are the respective time lags between task j and a previous task i. The FS relationship can be used to ensure continuity from one task to another by specifying $L_{FSij} = 0$. The relationships represented by Eqs. (5.6)–(5.9) can also be written for any given crew or segment of a task rather than the whole task. For example, consider the case where crew B of task Y is the same as crew A of a previous task X. Crew B can start work only after crew A has finished. Therefore, an FS relationship has to be specified between crew A of task X and crew B of task Y.

Step 7: Define the space and/or time buffer between tasks. These constraints are essential if interference of crews on different tasks is to be prevented. If task i precedes task j by a distance buffer B_{Sij}, the following constraints have to be satisfied:

$$Z_j(T_{in}^k) + B_{Sij} \leq Z_{in}^k \quad k = 1,\dots,N_{Ci}, n = 1,\dots,N_{Si}^k \qquad (5.10)$$

$$Z_j(T_{in}^k + D_{in}^k) + B_{Sij} \leq Z_{in}^{k'} \quad k = 1,\dots,N_{Ci}, n = 1,\dots,N_{Si}^k \qquad (5.11)$$

$$Z_{jn}^k + B_{Sij} \leq Z_i(T_{jn}^k) \quad k = 1,\dots,N_{Cj}, n = 1,\dots,N_{Sj}^k \qquad (5.12)$$

$$Z_{jn}^k + B_{Sij} \leq Z_i(T_{jn}^k + D_{jn}^k)' \quad k = 1,\dots,N_{Cj}, n = 1,\dots,N_{Sj}^k \qquad (5.13)$$

The term $Z_i(T_{jn}^k)$ denotes the location of task i at the time T_{jn}^k. For tasks with a constant production rate during a segment of work, $Z_i(T_{jn}^k)$ is found by a linear interpolation between the values at the start (Z_{im}^l) and the finish ($Z_{im}^{l'}$) of segment m performed by crew l of

task i.

$$Z_i(T_{jn}^k) = Z_{im}^l + \frac{(T_{jn}^k - T_{im}^l)(Z_{im}^{l'} - Z_{im}^l)}{(T_{im}^l + D_{im}^l) - T_{im}^l} \qquad (5.14)$$

Similarly, if task i precedes task j by the time buffer $B_{T_{ij}}$, then we

have the following constraints:

$$T_{in}^k + B_{T_{ij}} \le T_j(Z_{in}^k) \quad k = 1, \dots, N_{Ci}, n = 1, \dots, N_{Si}^k \qquad (5.15)$$

$$(T_{in}^k + D_{in}^k) + B_{T_{ij}} \le T_j(Z_{in}^{k'}) \quad k = 1, \dots, N_{Ci}, n = 1, \dots, N_{Si}^k \qquad (5.16)$$

$$T_i(Z_{jn}^k) + B_{T_{ij}} \le T_{jn}^k \quad k = 1, \dots, N_{Cj}, n = 1, \dots, N_{Sj}^k \qquad (5.17)$$

$$T_i(Z_{jn}^{k'}) + B_{T_{ij}} \le (T_{jn}^k + D_{jn}^k) \quad k = 1, \dots, N_{Cj}, n = 1, \dots, N_{Sj}^k \qquad (5.18)$$

Likewise, $T_i(Z_{jn}^k)$ is found by a linear interpolation between the

starting time (T_{im}^l) and the stopping time ($T_{im}^l + D_{im}^l$) for segment m

performed by crew l of task i.

$$T_i(Z_{jn}^k) = T_{im}^l + \frac{(Z_{jn}^k - Z_{im}^l)\{(T_{im}^l + D_{im}^l) - T_{im}^l\}}{(Z_{im}^{l'} - Z_{im}^l)} \qquad (5.19)$$

The optimization problem can now be formulated as the

minimization of direct cost

$$C_D = \sum_{i=1}^{N_{NT}} W_i C_i(d_i) + \sum_{i=1}^{N_{KI}} \sum_{k=1}^{N_{Ci}} \sum_{j=1}^{N_{Si}^k} W_{ij}^k C_i(d_i^k) \tag{5.20}$$

subject to the scheduling constraints [Eqs. (5.2)–(5.13) and (5.15)–(5.18)], plus initial constraint on the start time for the first segment of work performed by the first crew of the first task

$$T_{11}^1 = \text{const} \tag{5.21}$$

project duration constraints

$$T_{ij}^k + D_{ij}^k \leq D^{\max} \quad i = 1,\ldots,N_T, k = 1,\ldots,N_{Ci}, j = 1,\ldots,N_{Si}^k, \tag{5.22}$$

task duration constraints

$$(d_i^k)^{\min} \leq d_i^k \leq (d_i^k)^{\max} \quad i = 1,\ldots,N_T, k = 1,\ldots,N_{Ci}, \tag{5.23}$$

and, non-negativity constraints

$$T_{ij}^k, d_i^k \geq 0 \quad i = 1,\ldots,N_T, k = 1,\ldots,N_{Ci}, j = 1,\ldots,N_{Si}^k, \tag{5.24}$$

where C_i is the direct cost per unit quantity of work for task i; d_i^k is the time required by crew k of task i to complete a unit quantity of work based on resource allocation only; $(d_i^k)^{\min}$, $(d_i^k)^{\max}$ are the minimum and maximum possible values of d_i^k, respectively; and $D^{\max}$ is the maximum acceptable project duration. Note that in this formulation, Eq. (5.1) can be written for each crew k of task i as:

$$D_{ij}^k = \mu_{ij}^k d_i^k W_{ij}^k \tag{5.25}$$

5.4 CONCLUSION

In this chapter a general formulation was presented for the scheduling of construction projects. An optimization formulation was described for the construction project scheduling problem with the goal of minimizing the direct construction cost. In the next chapter, we adapt the neural dynamics model of Adeli and Park for the solution of the construction scheduling and cost optimization problem.

NEURAL DYNAMICS COST OPTIMIZATION MODEL FOR CONSTRUCTION PROJECTS

6.1   INTRODUCTION

In Chapter 5, we presented the general mathematical formulation for integrated scheduling and cost optimization of construction projects. In this chapter, we develop the neural dynamics construction scheduling and cost optimization model by adapting the general neural dynamics model of Adeli and Park (1995a) to the construction scheduling and cost optimization problem. A highway construction project example is then presented to illustrate the capabilities of the new model.

6.2   FORMULATION OF THE NEURAL DYNAMICS CONSTRUCTION COST OPTIMIZATION MODEL

Defining $\mathbf{X} = \{T_{ij}^k, d_i^k \mid i = 1, N_T, k = 1, N_{Ci}, j = 1, N_{Si}^k)$ as the vector of decision variables defined in Chapter 5, the optimization problem can be written as:

Minimize

$$C_D = f(\mathbf{X}) = \sum_{i=1}^{N_{NI}} W_i C_i(d_i) + \sum_{i=1}^{N_{RI}} \sum_{k=1}^{N_{Ci}} \sum_{j=1}^{N_{Si}^k} W_{ij}^k C_i(d_i^k) \qquad (6.1)$$

subject to inequality constraints

$$g_j(\mathbf{X}) \leq 0 \quad j = 1, \ldots, J \qquad (6.2)$$

and equality constraints

$$h_k(\mathbf{X}) = 0 \quad k = 1, \ldots, K \qquad (6.3)$$

where $g_j(\mathbf{X})$ is the jth inequality constraint function, $h_k(\mathbf{X})$ is the kth equality constraint function, J is the total number of inequality constraints, and K is the total number of equality constraints. Using the exterior penalty function method, a pseudo-objective function is defined as:

$$P(\mathbf{X}, r_n) = f(\mathbf{X}) + \frac{r_n}{2} \left\{ \sum_{j=1}^{J} \left[g_j^+(\mathbf{X}) \right]^2 + \sum_{k=1}^{K} \left[h_k(\mathbf{X}) \right]^2 \right\} \qquad (6.4)$$

where $g_j^+(\mathbf{X}) = \max\{0, g_j(\mathbf{X})\}$ and r_n is a penalty parameter magnifying constraint violations.

A dynamic system is defined as:

$$\frac{d\mathbf{X}}{dt} = \dot{\mathbf{X}} = F(\mathbf{X}) \qquad (6.5)$$

where $\mathbf{X} = \{X_1(t), X_2(t), \ldots, X_N(t)\}^T$ is the state vector tracing a trajectory in N-dimensional space, the superscript T indicates the transpose of a vector and $N = \sum_{i=1}^{N_I} \sum_{k=1}^{N_{Ci}} N_{Si}^k + \sum_{i=1}^{N_T} N_{Ci}$. The dynamic

system evolves until it reaches an equilibrium point. The stability of

such an equilibrium point is ensured by satisfying the Lyapunov

stability theorem which states that a solution $\mathbf{X}$ to the system of

differential equations $\dot{\mathbf{X}} = 0$ is stable if

$$\frac{dV}{dt} \le 0 \quad \text{for all non-zero } \mathbf{X} \tag{6.6}$$

where $V(\mathbf{X})$ is the Lyapunov functional defined as an analytic

function of the state variables such that $V(\mathbf{0}) = 0$ and $V(\mathbf{X}) > 0$ for

all $|\mathbf{X}| > 0$ (Kolk and Lerman, 1992). The objective (direct cost)

function and the constraint functions in our construction cost

optimization model both satisfy the conditions for a Lyapunov

functional. Therefore, the pseudo-objective function P defined by

Eq. (6.4) is also a valid Lyapunov functional, V.

Following Adeli and Park (1995b), by defining

$$\frac{d\mathbf{X}}{dt} = \dot{\mathbf{X}} = -\nabla f(\mathbf{X}) - r_n \left\{ \sum_{j=1}^{J} g_j^+ \nabla g_j(\mathbf{X}) + \sum_{k=1}^{K} h_k \nabla h_k(\mathbf{X}) \right\} \tag{6.7}$$

where $\nabla f(\mathbf{X})$, $\nabla g_j(\mathbf{X})$, and $\nabla h_k(\mathbf{X})$ are the gradients of the

objective function, the jth inequality constraint, and kth equality

constraint, respectively, the Lyapunov stability theorem for the

dynamic system is satisfied.

$$\frac{dV}{dt} = \left(\frac{dV}{d\mathbf{X}}\right)\left(\frac{d\mathbf{X}}{dt}\right)$$

$$= -\left[\nabla f(\mathbf{X}) + r_n \left\{\sum_{j=1}^{J} g_j^+ \nabla g_j(\mathbf{X}) + \sum_{k=1}^{K} h_k \nabla h_k(\mathbf{X})\right\}\right]^2 \leq 0 \qquad (6.8)$$

This also shows that the dynamic system evolves such that the value of the pseudo-objective function always decreases. Equation (6.7) is in fact the learning rule of the neural dynamics model.

For an equilibrium point $\mathbf{X}$ to be a local optimum solution, we also need to satisfy the Kuhn-Tucker optimality conditions:

$$\frac{\partial L}{\partial X_i} = \frac{\partial f(\mathbf{X})}{\partial X_i} + \sum_{j=1}^{J} u_j \frac{\partial g_j(\mathbf{X})}{\partial X_i}$$

$$+ \sum_{k=1}^{K} v_k \frac{\partial h_k(\mathbf{X})}{\partial X_i} = 0; \quad i = 1,\dots,N \qquad (6.9)$$

$$g_j(\mathbf{X}) + s_j^2 = 0; \quad j = 1,\dots,J \qquad (6.10)$$

$$h_k(\mathbf{X}) = 0; \quad k = 1,\dots,K \qquad (6.11)$$

$$u_j s_j = 0; \quad j = 1,\dots,J \qquad (6.12)$$

$$u_j \geq 0; \quad j = 1,\dots,J \qquad (6.13)$$

$$v_k = \text{unrestricted in sign} \qquad (6.14)$$

where L is the Lagrangian function defined as a linear combination of the objective and constraint functions:

$$L(\mathbf{X}, \mathbf{u}, \mathbf{v}, \mathbf{s}) = f(\mathbf{X}) + \sum_{j=1}^{J} u_j \left[g_j(\mathbf{X}) + s_j^2 \right] + \sum_{k=1}^{K} v_k h_k(\mathbf{X}), \qquad (6.15)$$

in which s_j is the slack term for the jth inequality constraint, and u_j and v_k are the Lagrangian multipliers corresponding to the jth inequality and kth equality constraint, respectively.

Finally, the optimum solution to the direct cost optimization problem can be found by the integration:

$$\mathbf{X} = \int \dot{\mathbf{X}} dt. \qquad (6.16)$$

This integration can be performed by the Euler or Runge-Kutta method.

6.3   TOPOLOGICAL CHARACTERISTICS

The neural network topology for the neural dynamics construction cost optimization model is shown in Figure 6.1. The nodes in the network represent the variables and constraints of the problem. The variable layer has N nodes corresponding to the total number of decision variables. The constraint nodes are divided into N_{NT} layers corresponding to non-repetitive tasks, N_{RT} layers corresponding to repetitive tasks, and an initial constraint node. Nodes are grouped within each layer into the constraint categories described in the previous   chapter.   Variable   and   constraint   nodes   are   fully

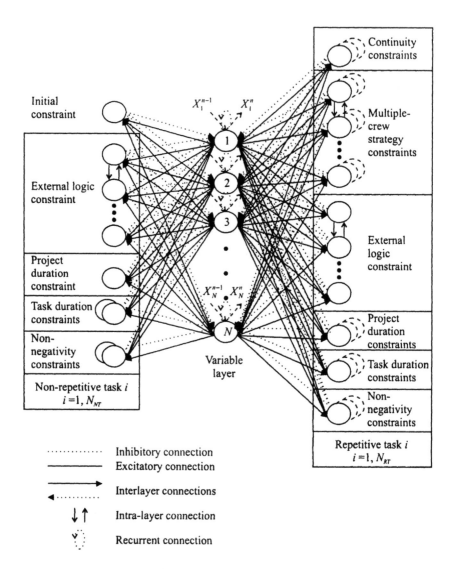

Figure 6.1   Neural network topology for neural dynamics cost optimization model

interconnected (interlayer connections). In addition, recurrent and intra-layer connections are also used, to be described shortly.

Associated with each connection is a weight whose magnitude and sign affect the impulse the connected node will receive. Both excitatory (positive connection weights) and inhibitory (negative connection weights) connections are used in our model. The coefficients of the constraint functions are assigned to the excitatory connections from the variable layer to the constraint nodes. The gradients of the constraint functions are assigned to the inhibitory connections from the constraint nodes to the variable layer. The gradients of the objective function are assigned to the recurrent inhibitory connections of the variable layer. A constant weight of one is assigned to the intra-layer connections between constraint nodes within a constraint category. This allows the outputs of nodes in each constraint category to be compared.

The output of the variable layer is the current state vector $\mathbf{X}$. As the coefficients of the constraint functions are encoded in the excitatory connections from the variable layer to the constraint nodes, the input to a constraint node is the magnitude of the constraint at any given state, that is, $g_j(\mathbf{X})$ for an inequality constraint j, and $h_k(\mathbf{X})$ for an equality constraint k. The output of a constraint node will depend on the type of the constraint it represents. For an inequality constraint j, the output is:

$$O_{cj} = \begin{cases} 0 & \text{when } g_j(\mathbf{X}) \leq 0 \\ r_n g_j(\mathbf{X}) & \text{when } g_j(\mathbf{X}) > 0 \end{cases} \tag{6.17}$$

and for an equality constraint k the output is:

$$O_{ck} = \begin{cases} 0 & \text{when } h_k(\mathbf{X}) = 0 \\ r_n h_k(\mathbf{X}) & \text{when } h_k(\mathbf{X}) \neq 0 \end{cases} \tag{6.18}$$

Equations (6.17) and (6.18) represent the activation functions. They are chosen such that the output of a constraint node is the penalized constraint violation. When more than one equation is specified for a particular category of constraint, such as external logic constraint, a competition is created between the outputs of the nodes in that group. For a group of n nodes within a constraint category with outputs (given by Eqs. 6.17 and 6.18) $O_{c1}, O_{c2}, \ldots, O_{cj}, \ldots, O_{cn}$ such that

$$O_{cj} = \max\{O_{c1}, O_{c2}, \ldots, O_{cj}, \ldots, O_{cn}\} \tag{6.19}$$

The outputs after competition are O_{cj} as defined by Eq. (6.19) and

$$O_{c1}, O_{c2}, \ldots, O_{cn} = 0 \quad \text{(excluding } O_{cj}) \tag{6.20}$$

In words, after competition the outputs of all nodes become zero except for the node with the largest output.

Let w_{ji} and w_{ki} be the weights of the links connecting the jth and kth inequality and equality constraint node, respectively, to the ith variable node and Y_i be the weight of the recurrent connection to a node i in the variable layer. Then, the input to the ith variable node is given by:

$$I_{vi} = Y_i + \sum_{j=1}^{J} w_{ji} O_{cj} + \sum_{k=1}^{K} w_{ki} O_{ck} \qquad (6.21)$$

The new value of the ith decision variable is obtained by the integration:

$$X_i^{new} = \int I_{vi} dt \qquad (6.22)$$

This integration is done by the Euler or the Runge-Kutta method. In the construction cost optimization problem we found the simple Euler method to yield accurate results.

The network operates until no change in the decision variables occurs within a given tolerance, that is, when $\dot{X} = F(X) = 0$ within a given tolerance. X is then the solution to the minimum direct cost construction scheduling problem.

6.4   ILLUSTRATIVE EXAMPLE

6.4.1   General Description

A 5 km-long two-lane highway construction project (Figure 6.2) is used to illustrate the capabilities of the computational model presented. The work required is divided into 14 repetitive and non-repetitive tasks summarized in Table 6.1. Tasks 1 to 5 represent the establishment of a temporary site office at the beginning of the 5 km

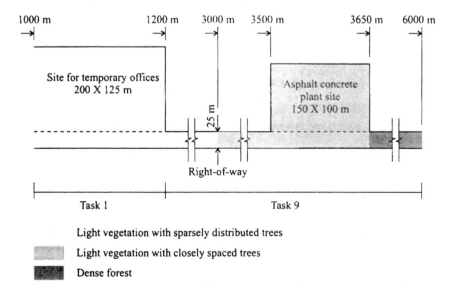

Figure 6.2   Areas and types of vegetation that have to be cleared by tasks 1 and 9 for illustrative example

Table 6.1   Description and type of tasks for illustrative example

Task #	Description	Type
1	Clear and grub site for temporary offices plus right-of-way	Non-repetitive
2	Grade site for temporary offices	Non-repetitive
3	Erect temporary offices	Non-repetitive
4	Construct temporary roads	Non-repetitive
5	Move-in	Non-repetitive
6	Grade asphalt concrete plant site	Non-repetitive
7	Erect asphalt concrete plant	Non-repetitive
8	Construct culverts	Repetitive
9	Clear and grub right-of-way	Repetitive
10	Earthwork	Repetitive
11	Lay sub-base	Repetitive
12	Lay base	Repetitive
13	Pave	Repetitive
14	Finish shoulders	Repetitive

Table 6.2   Direct cost-duration relationship for each task

Task #	Direct cost-duration relationship	Range (days)
1	$C = -300d + 1050$	$1.0 \leq d \leq 1.5$
2	$C = -280d + 960$	$0.5 \leq d \leq 2.0$
3	$C = -200d + 250$	$0.25 \leq d \leq 0.50$
4	$C = -200d + 550$	$0.50 \leq d \leq 1.25$
5	$C = -150d + 550$	$1.0 \leq d \leq 2.0$
6	$C = -280d + 960$	$0.5 \leq d \leq 2.0$
7	$C = -400d + 5700$	$5 \leq d \leq 8$
8	$C = 1600/d$	$2 \leq d \leq 3$
9	$C = -300d + 1050$	$1.0 \leq d \leq 1.5$
10	$C = (1600 + 500d)/d$	$1.0 \leq d \leq 2.0$
11	$C = -200d + 850$	$0.75 \leq d \leq 1.25$
12	$C = -200d + 950$	$0.75 \leq d \leq 1.25$
13	$C = -200d + 900$	$0.75 \leq d \leq 1.25$
14	$C = -100d + 800$	$2 \leq d \leq 4$

long stretch. The erection of an asphalt concrete plant at a distance of 2.5 km from the beginning of the roadway (at the center of the project) is represented by tasks 6 and 7 (together with portion of task 9).

6.4.2   Cost-Duration Relationship

The relationship between direct cost and duration for unit quantity of work for each task is given in Table 6.2. An initial cost of $5000 and, thereafter, a daily cost of $500 is used as the indirect cost for this example.

6.4.3   Scheduling Logic

The way in which each task is performed and the logic in which the tasks are carried out for a given project are not always well defined.

Different schedulers may have different ideas for breaking down and sequencing each task. Often schedulers are constrained by the scheduling model available, forcing them to make simplifying assumptions. The flexible computational models presented in Chapters 5 and 6, however, allows schedulers a greater control over the progress of work and enables them to complete the job more efficiently.

Details of the breakdown of repetitive tasks into crews and segments, the start and finish distances, the quantities of work required, and the job condition factors for segments of work are given in Table 6.3. A constant number (1000 m) is used as the start distance of the project to avoid division by zero in the computation.

How the variation in the quantities of work and the job condition factors affect the breakdown of tasks can be explained by the clear and grub operations represented by tasks 1 and 9. Figure 6.2 shows the areas that have to be cleared and grubbed and the type of vegetation involved. Task 1 operates on the first 200 m of the roadway including the area for the site office. Task 9 covers the remaining length of the highway including the site for the asphalt concrete plant. A new segment of work is required whenever there is a change in the quantity of work required per unit length of the highway and/or a change in the job condition factor. Each change will affect the production rate. To reflect such a change a separate segment of work is defined. Whenever there is no such change, such

Table 6.3 Task details for illustrative example

Task #	Crew #	Segment #	Distances (m) Start	Distances (m) Finish	Quantity of work	Unit	Job condition factor
1	1	-	1000	1200	3	hectares	1.0
2	1	-	1000	1200	3	hectares	1.0
3	1	-	1000	1200	5	units	1.0
4	1	-	1000	1200	3	100 m	1.0
5	1	-	1000	1200	100	percent	1.0
6	1	-	3500	3650	1.5	hectares	1.0
7	1	-	3500	3650	100	percent	1.0
8	1	1	1300	1305	1	culvert	1.0
		2	2750	2755	1	culvert	1.0
		3	5500	5505	1	culvert	1.0
9	1	1	1200	3000	4.5	hectares	1.0
		2	3000	3500	1.25	hectares	1.1
		3	3500	3650	1.875	hectares	1.1
		4	3650	6000	5.875	hectares	1.15

Table 6.3 – continued

Task #	Crew #	Segment #	Distances (m) Start	Distances (m) Finish	Quantity of work	Unit	Job condition factor
10	1	1	1000	3000	10	1000 m³	1.0
		2	3000	3500	6	1000 m³	1.15
	2	1	3500	5000	8	1000 m³	1.05
		2	5000	6000	5	1000 m³	1.0
11	1	1	1000	2000	4.25	1000 m³	1.0
		2	2000	4000	8.5	1000 m³	1.05
		3	4000	6000	8.5	1000 m³	1.0
12	1	1	3500	2250	2.6	1000 m³	1.0
		2	2250	1000	2.6	1000 m³	1.05
	2	1	3500	4750	2.6	1000 m³	1.0
		2	4750	6000	2.6	1000 m³	1.05
13	1	1	3500	2250	1.25	1000 m³	1.0
		2	3250	1000	1.25	1000 m³	1.05
	2	1	3500	4750	1.25	1000 m³	1.0
		2	4750	6000	1.25	1000 m³	1.05
14	1	1	1000	6000	3	hectares	1.0

as for task 14, there is no need to break down the work into smaller segments.

Base laying and paving operations (tasks 12 and 13) require material from the asphalt concrete plant. Therefore, as the operation moves away from the plant more time will be taken to do the same amount of work. In our example, we increased the time required for operations beyond 1250 m from the plant by 5 percent indicated by the job condition factor of 1.05. Instead of a step function, a continuous linear or nonlinear function may be used for the job condition factor that will reflect the impact of increasing haul distances on the rate of operation.

The internal logic of repetitive tasks is given in Table 6.4. Work continuity relationships between segments of work and multiple-crew strategies are specified. Usually no slack time ($S_{ij}^k = 0$) is allowed between segments of the work of a crew. However, the slack term can be any function of the decision variables. We use a nonlinear slack term to model the continuity of work constraint of task 8 in the form:

$$S = 1 - \beta \le 1.0 \tag{6.23}$$

where $\beta < 1.0$ is the fractional portion of the finishing time (starting time plus duration) of the previous segment of work. For example, for a finishing time of 10 days and 2 hours (10.25 days assuming 8 hours per day) $\beta = 0.25$. This insures that the work on the next segment will start on the following day. As a result, adequate time is

Table 6.4   Internal logic of repetitive tasks

Task #	Crew #	Segment #	Continuity relationship	Multiple-crew strategy	
			$-$	Predecessor crew	Relationship
8	1	1	$-$	$-$	$-$
		2	$S = 1 - \beta^{a}$		
		3	$S = 1 - \beta^{a}$		
9	1	1	$-$	$-$	$-$
		2	$S = 0$		
		3	$S = 0$		
		4	$S = 0$	$-$	$-$
10	1	1	$-$		
		2	$S = 0$	1	SS, $L = 0$
	2	1	$-$		
		2	$S = 0$		
11	1	1	$-$	$-$	$-$
		2	$S = 0$		
		3	$S = 0$		
12	1	1	$-$	$-$	$-$
		2	$S = 0$		
	2	1	$-$	1	FF, $L = 0$
		2	$S = 0$		
13	1	1	$-$	$-$	$-$
		2	$S = 0$		
	2	1	$-$	1	SS, $L = 0$
		2	$S = 0$		

[a] β is the fractional portion of the finishing time of the previous segment of work

Table 6.5   External logic of tasks

Task	Predecessor	Relationship
Task 1		
Task 2	Task 1	FS, $L = 0$
Task 3	Task 2	SS, $L = 0$
	Task 4	FS, $L = 0$
Task 4	Task 2	FS, $L = 0.25D$
Task 5	Task 3	FS, $L = 0$
Task 6	Segment 3, Crew 1, Task 9	FS, $L = 0$
Task 7	Task 6	FS, $L = 0$
Task 8: Crew 1, Segment 1	Task 9 at 1300 m	FS, $L = 0$
Crew 1, Segment 2	Task 9 at 2750 m	FS, $L = 0$
Crew 1, Segment 3	Task 9 at 5500 m	FS, $L = 0$
Task 9	Task 1	FS, $L = 0$
Task 10	Task 9	Space buffer, $B = 150$ m
Task 11	Task 10	Space buffer, $B = 150$ m
Task 12	Task 7	FS, $L = 0$
	Task 11	Time buffer, $B = 2$ days
Task 13	Task 12	Time buffer, $B = 2$ days
Task 14	Task 13	Time buffer, $B = 2$ days

provided for the crew to move from one location to the next. Multiple-crew strategies become important when more than two crews are used.

Table 6.5 gives the external logic of tasks for the illustrative example. The external logic of the first 5 tasks, which are non-repetitive, can also be shown by an activity-on-node (AON) diagram (Figure 6.3). Standard precedence relationships are used to link the tasks. The time lag term, however, may be any function of the decision variables.

The construction of a culvert cannot start unless the area has been cleared and grubbed. Therefore, the external logic of task 8 requires that work on any culvert be delayed until the crews of repetitive

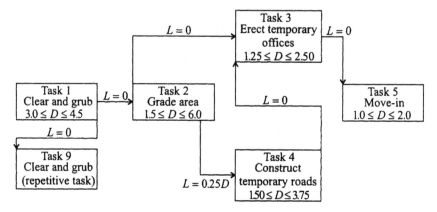

D = Duration in days
L = Lag in days

Figure 6.3   Activity-on-node   diagram   for   first   five   tasks   of illustrative example

task 9 have worked through the corresponding location. A space buffer of 150 m is provided around earthmoving operations (task 10) to make sure adequate space is available for the equipment. Tasks 12 and 13 cannot start before the completion of the asphalt concrete plant. A minimum time buffer of 2 days is provided between tasks 11, 12, 13 and 14.

6.4.4   Solution of the Problem

The direct cost optimization problem is solved for project durations of 60, 65, 70, 80, 90 and 100 days. Following Adeli and Park (1995a) the penalty parameter, r_n, is taken as:

$$r_n = r_0 + \frac{n}{\alpha} \qquad\qquad (6.24)$$

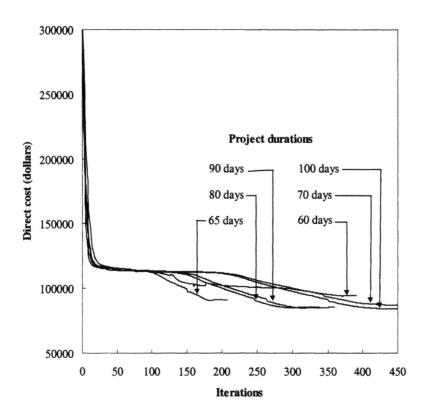

Figure 6.4   Direct cost convergence curves for illustrative example

where r_0 is the initial penalty, n is the iteration number and α is a positive number. Through this relationship the penalty is increased gradually in each iteration to avoid the possibility of numerical ill-conditioning. As stopping criteria, a change of less than $1 in the original objective (direct cost) function and a maximum of 450 iterations are chosen. The convergence curves for the solutions are given in Figure 6.4. Table 6.6 and Figure 6.5 show the variation of

Table 6.6   Direct, indirect, and total costs variation for illustrative example

Duration (days)	Direct cost (dollars)	Indirect cost (dollars)	Total cost (dollars)
60	94118	30000	124118
65	91215	32500	123715
70	87314	35000	122414
80	85742	40000	125742
90	85438	45000	130438
100	84411	50000	134411

direct, indirect, and total costs for different values of project duration. From Figure 6.5 and Table 6.6, a project duration of 70 days leads to the minimum total cost. The final global optimum schedule is shown as a linear planning chart in Figure 6.6.

6.5   CONCLUSION

In Chapter 5, a general formulation was presented for the scheduling of construction projects. Both repetitive and non-repetitive tasks are considered in the formulation. By specifying appropriate constraints, work continuity considerations and multiple-crew strategies can be modeled. The effects of varying job conditions on the performance of a crew is taken into account by introducing a job conditions factor which modifies the task duration computed on the basis of resource allocation only. This factor can be a constant, a linear, or a nonlinear function depending on the complexity of the situation. An optimization formulation is presented for the construction project scheduling problem with the goal of minimizing the direct

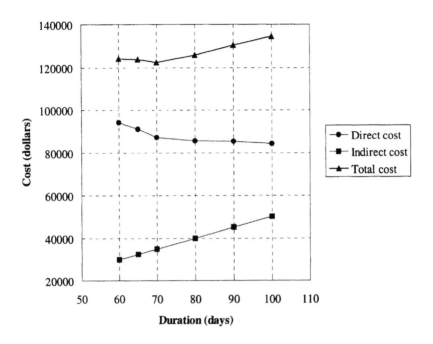

Figure 6.5   Time-cost trade-off curve for illustrative example

construction cost. Any linear or non-linear function can be used for task direct cost-duration relationships.

In this chapter, the nonlinear optimization problem was solved by the neural dynamics model developed recently by Adeli and Park (1995a). For any given construction duration, the model yields the optimum construction schedule for the minimum construction direct cost automatically. By varying this construction duration, one can solve the cost-duration trade-off problem and obtain the global optimum schedule and the corresponding minimum construction

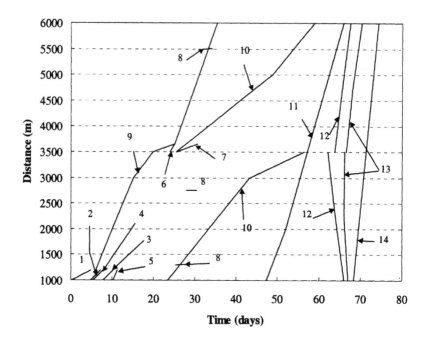

Figure 6.6   Linear planning chart for minimum total cost schedule

cost. A highway construction project example was presented to illustrate the capabilities of the developed model.

The new construction scheduling model provides the capabilities of both CPM and LSM approaches, In addition, it provides features desirable for repetitive tasks such as highway construction projects and allows schedulers greater flexibility in modeling construction projects more accurately. In particular, it is suitable for studying the effects of change order on the construction cost. The new scheduling model can be specialized for the solution of specific and perhaps less complicated scheduling problems.

7

OBJECT-ORIENTED INFORMATION MODEL FOR CONSTRUCTION PROJECT MANAGEMENT

7.1   INTRODUCTION

In Chapter 5 we developed a general mathematical model for scheduling of the construction projects. The model can handle various conditions such as repetitive and non-repetitive tasks, work continuity considerations, multiple-crew strategies, and the effects of varying job conditions on the performance of a crew. An optimization formulation was presented for the construction project scheduling problem with the goal of minimizing the direct construction cost. In Chapter 6, the nonlinear optimization problem was solved by the neural dynamics model of Adeli and Park.

In this chapter an object-oriented information model is presented for construction scheduling, cost optimization, and change order management based on the new construction scheduling model presented in previous two chapters. The model can be used by the owner/client who has to approve any change order requests made by the contractor, as well as by the contractor. The model provides

177

support for schedule generation and review, cost estimation, and cost-time trade-off analysis. Furthermore, the model implemented in a software system can be used as an intelligent decision support system to resolve change order conflicts.

7.2   CHANGE ORDER MANAGEMENT

Change in construction projects after bid are a common occurrence. Change orders often lead to disputes between the owner and the contractor. These disputes are usually on the issue of compensation for change. The lack of a consistent method of change order management can result in long delays and costly legal battles.

Broadly, change in a construction project after bid may be required because of the following reasons (Saunders, 1996):

- An error or inconsistency in the original specification or design.
- A change in the functional requirement of the project after bid.
- An unforeseen or unknown condition requiring a change in the original construction method and/or design.

Computer support for change order management, as reported in the literature, is minimal. Leymeister et al. (1993) describe the use of databases to analyze and process contractors' direct cost payment requests. Their goal is to reduce the possibility of errors, omissions, and duplication of claims made by the contractor. No support is provided for determining the validity of the claims and/or their impact on the project. De Leon and Knoke (1995) present a probabilistic procedure based on the critical path method (CPM) to determine the extension needed for contract time as a result of changes. Impact of changes on cost is not considered. Other recent publications on construction change order management include a survey of construction cost markups paid on direct cost as a result of changes (Saunders, 1996), a study of changes and their impact based on the survey of private sector projects (Ibbs, 1997), and classification, identification, and analysis of factors causing delays by the contractor (Abd. Majid and McCaffer, 1998). These articles show the lack of a consistent and sufficiently formal procedure for handling change orders in construction. In most cases the change orders are handled somewhat arbitrarily.

7.3 OWNER'S ROLE IN CONSTRUCTION PROJECT MANAGEMENT

The majority of major construction projects are awarded by the government agencies such as state Departments of Transportation (DOT). These agencies are involved throughout the life-cycle of a project from the conception of the project to its maintenance and operation. The owner's role from project conception to final award of the contract and the phases involved are shown in Figure 7.1. The only significant change that has to be handled in this stage is that of scope change after work has started on the detailed design and planning phase. Such changes can be handled relatively easily and without significant additional costs to the project.

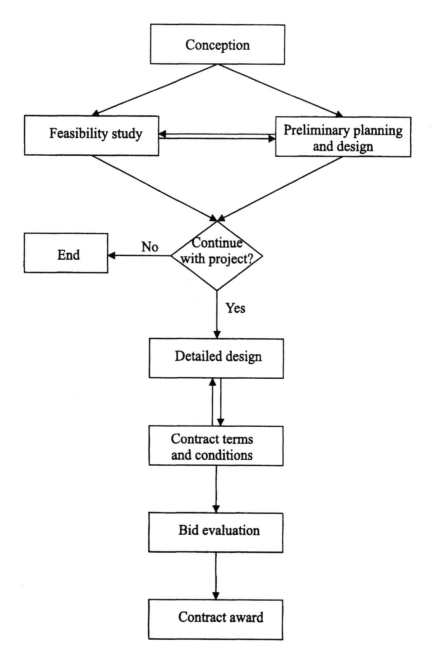

Figure 7.1   Owner's role from project conception to final project award

After the contract is awarded, the owner is responsible for progress monitoring, schedule reviews, change order management, and conflict management. Any change occurring during this stage of the project life-cycle is bound to have a profound impact on the project cost and duration. Furthermore, any such change has a potential to create a conflict between the contractor and the owner on issues such as the compensation for change. The major phases in change order management from the owner's perspective are presented in Figure 7.2.

The owner-contractor interaction during the change order involves a lot of information exchange. The owner has to analyze this information carefully and make intelligent decisions in a timely manner. An owner, such as a state DOT, has to protect its interests while at the same time be fair to the contractor. A flexible computer support system for change order management can provide efficiency, consistency, reliability, and credibility to the decision making process.

7.4   OBJECT-ORIENTED METHODOLOGY AND CONSTRUCTION ENGINEERING

Use of the object technology has gained increasing popularity in the development of flexible, maintainable, and reusable software systems (Yu and Adeli, 1991, 1993; Adeli and Yu, 1993a,b; Adeli and Kao, 1996; Kao and Adeli, 1997). The basic concept of object-orientation is the object that abstracts a real-world entity by

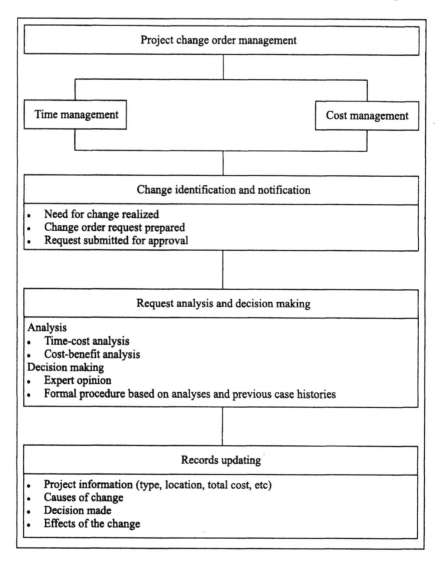

Figure 7.2   Project change order management (owner's perspective)

encapsulating its characteristics (data and functionality). An object provides an interface for communication with other objects.

Constructs such as inheritance and polymorphism allow easy extension and reusability of previously developed objects. The object-oriented methodology provides an information processing paradigm for efficient development and management of complicated software systems.

The new scheduling/cost optimization model presented in the previous two chapters is more general and flexible and provides the mathematical foundations for development of a new generation of construction scheduling software systems. The computational model is compatible with CPM but provides additional features like multiple-crew support, separate constructs for repetitive and non-repetitive tasks, time and space buffer constraints, and the capability to model distance/locations of tasks in scheduling. Further, it provides a robust cost optimization capability that can handle both linear and nonlinear cost-duration relationships. Cost optimization of construction projects reported in the literature is limited to CPM or CPM-like models of the project (Liu *et al.*, 1995; Feng *et al.*, 1997). Such models cannot model various construction projects such as highway construction accurately.

Use of the object-oriented methodology in computer-integrated construction (CIC) has been reported in the recent literature. However, most of this research is on developing standard project information models to support concurrent engineering in the architecture, engineering, and construction (AEC) industry (Fischer and Froese, 1996; Froese, 1996; Stumpf *et al.*, 1996). A common characteristic of these models is their emphasis on domain modeling with little discussion on computational modeling. Froese and Paulson (1994) present an object model-based project information system as a standard information model for the AEC industry. System integration is achieved by a shared object database. A construction scheduling application module is tested as an example. Fischer and Aalami (1996) advocate the use of construction method knowledge in the generation of construction schedules. They describe object models of the construction methods that can dynamically link construction design and scheduling information.

The object-oriented information model presented in this chapter for construction scheduling, cost optimization, and change order management is implemented as an application development *framework* in Visual C++. The use of *framework* allows software design to be encoded in a reusable format for rapid development of compatible software systems. As an example, the object oriented framework is used to develop an intelligent decision making tool that can be used by the owner in its dealings with a contractor. The object-oriented model can also be an important part of a concurrent engineering model for the AEC industry.

7.5   AN OBJECT-BASED INFORMATION MODEL FOR CONSTRUCTION SCHEDULING, COST OPTIMIZATION, AND CHANGE ORDER MANAGEMENT

How do we arrange objects to solve a particular problem keeping in mind the goals of effective reusability, maintainability, and extensibility for complex software systems? A number of different software designs and architectures have been proposed for this fundamental software engineering problem. Broadly speaking, software design involves developing architecture, techniques, and strategies for data handling and manipulation, user-interface design, and program control. The software design decision may be a major one like whether to use a relational or an OO database management system, or it may be minor like what data type to use for a particular kind of data. A software design approach with desirable aforementioned characteristics is the *framework*-based design.

A framework is a collection of interacting and cooperating software components for solving a generic category of tasks such as database management or user interface design and/or a domain specific task such as construction scheduling. In an OOP environment, the components can be connected through various constructs such as object references or pointers. Software design decisions such as the construct to use to allow cooperation between two components are captured by a framework in a reusable format. In addition to the advantages gained by software reuse in general such as reduction in development time and cost and increase in software reliability, encapsulating an OO model in a framework leads to the following benefits:

- The OO model does not have to be reinterpreted and coded whenever a new software application in the domain is developed.

- The use of frameworks leads to compatible software systems because each shares the same underlying software design.

7.6   SOFTWARE REUSE TECHNIQUES: COMPONENTS, DESIGN PATTERNS, AND FRAMEWORKS

Reuse is a major concern in creating significant software systems. Effective reuse techniques can reduce development and maintenance costs substantially. Reuse in software engineering is not limited to code only. But rather, reuse encompasses information and knowledge that has been gained over time including software development methods and designs that have been tested and proven efficient (Jacobson *et al.*, 1992). Object-oriented technology provides a reusable software environment that can take advantage of both design and code reuse. This is a primary advantage of the OO technology.

The most common form of reuse is through the use of *components*. Components can be regarded as software building blocks. As such, they raise the level of abstraction for the software developer who no longer has to worry about the low level implementation details. Examples of components are subroutines in

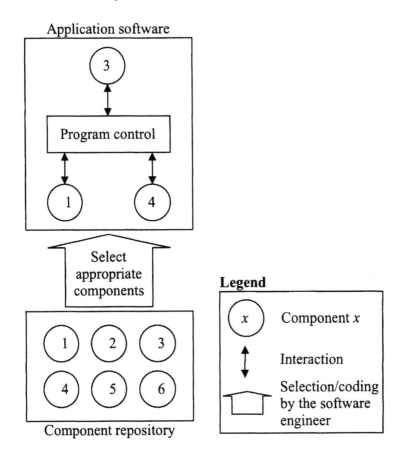

Figure 7.3   Software reuse using components

FORTRAN and classes in C++. Software development using components involves coding the application architecture and control and plugging in components whenever an appropriate one is available (Figure 7.3). This approach to software reuse has a number of limitations including:

12. Components reuse only code;

13. Components are not very flexible, that is, they are good at performing one particular task only and cannot be customized; and

14. Components require a detailed and problem-specific archiving plan.

Even though components can be developed in an OOP language their use requires a bottom-up approach, which is not desirable for object-oriented software development. The concept of component reuse is shown in Figure 7.3. A major amount of programming is still required for application development including lower level details; components just fill in for specialized tasks.

A *pattern* is a reusable software design (Gamma *et al.*, 1995). Patterns abstract commonly used tried and tested approaches in a descriptive format. The description includes the problem statement, the motivation for and the intent of the solution, the solution, the context in which the solution works, cost/benefit information, and important implementation details and tips. The solution is usually presented in a graphical manner that shows the static and dynamic relationships between the classes involved. A pattern usually provides descriptive information. Whenever a pattern is used it has to be coded in a programming language (Figure 7.4). But this, in turn, makes patterns platform independent and widely applicable. Few collections of patterns are available in the current literature. Gamma *et al.* (1995) catalog general-purpose commonly recurring software design patterns in software systems. More recently, Fowler (1997)

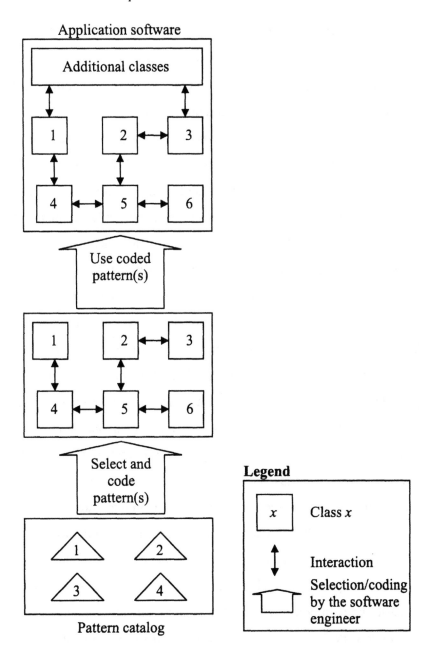

Figure 7.4   Software reuse using patterns

surveys software design patterns in the business, finance, and health care industry software systems. When widely recognized and adopted, design patterns can improve software reliability and facilitate design communication and understanding.

Framework, an important concept in object-oriented reuse technology (Johnson, 1997; Demeyer *et al.*, 1997, Rogers, 1997), consists of a set of inter-related abstract and concrete classes that form the skeleton of an application in a particular domain. A framework reuses software designs because it abstracts patterns and other software design decisions. But unlike patterns, frameworks are expressed in code. They are therefore simple or partially complete applications with common functionality and built-in control. Software development using frameworks involves customizing the frameworks for the particular application. One way to do this is by providing implementations for the abstract classes (Figure 7.5). A software application may make use of more than one framework. The use of multiple frameworks may be required when, for example, one framework depends on another for its working. Dependencies are usually specified in the overall application architecture.

The framework approach to reuse in software engineering is more flexible and extensible and hence more powerful than reuse through components. It also requires a lesser amount of coding because the common lower level details are already implemented in the framework. Further, the use of frameworks results in compatible, manageable, and extensible software systems. A significant

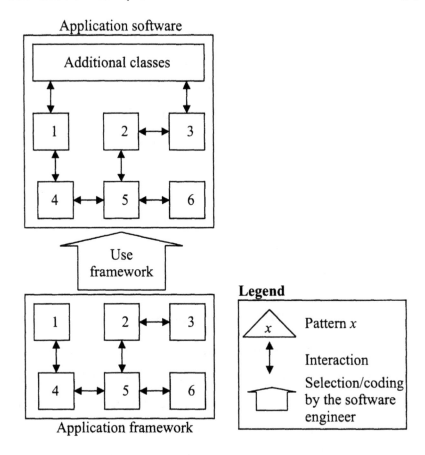

Figure 7.5   Software reuse using frameworks

distinction between application development with components and that with frameworks is *inversion of control* (Gamma *et al.*, 1995). In a framework-based software system, the developer writes code that is called by the framework. In contrast, the developer is responsible for calls to components in a component-based approach. On the downside, practical framework design is iterative and time consuming. Its use also requires good documentation and experience on the part of the developer (Schmidt and Fayad, 1997). However, these limitations are not unique to frameworks but apply to all OO reuse techniques. Fichman and Kemerer (1997) present case studies of OO reuse problems and lessons learned. Typical examples of frameworks for generic tasks are the C++ Standard Template Library (STL) and the Microsoft Foundation Class (MFC) library (Stepanov and Lee, 1995; Microsoft, 1997).

Karim and Adeli (1999a) appears to be the first article on the development and use of domain-specific software frameworks in civil engineering.

7.7   DEVELOPMENT ENVIRONMENT

A construction management application development framework is presented using Microsoft Foundation Class (MFC) library (Microsoft, 1997). For object-oriented programming in the Windows environment MFC is fast becoming the standard for software development. MFC is a general-purpose application development framework. The set of classes it provides abstracts the design and functionality of a simple Windows application. An actual application is built by reusing and adding to MFC. MFC provides basic application software design and control support, user-interface support, and limited file storage/retrieval support. Further, MFC provides general support classes for data handling and manipulation, and foundation classes for database management. A simplified class diagram of the MFC library is shown in Figure 7.6. Note that in the

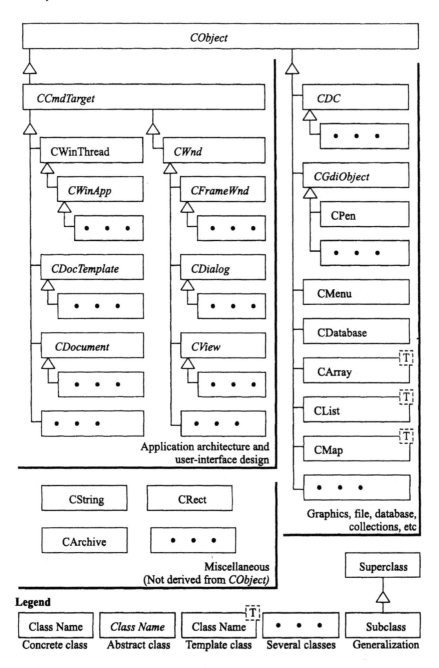

Figure 7.6   Simplified class diagram of MFC library

MFC convention, class names start with a capital C. The notation in this and all the subsequent figures is based on the new standard Unified Modeling Language (UML) (Fowler and Scott, 1997). Classes identified in Figure 7.6 are defined at the end of this chapter.

The MFC library supports the single document, multiple view concept of application design and control. The basic idea in this OO concept is the separation of the application data (the document) from its visual presentation (the view). A document object handles data processing, manipulation, storage, and integrity while data presentation and user feedback control are managed by the view objects. This is implemented by objects of classes derived from *CDocTemplate*, *CDocument*, and *CView* (Figure 7.6). These are abstract classes and cannot be instantiated. Application program control is provided by means of message maps which route Windows and user generated messages to the appropriate handling function. For a class to be able to receive messages, it must be derived from *CCmdTarget* (Figure 7.6). This is also an abstract or virtual class. It defines the interface for message processing. Each application must have a single global instance of the class *CWinApp* (Figure 7.6). This class encapsulates the execution of an application.

The user interface support classes abstract most of the common window elements and their functionality. Classes in this category include those for frame windows, views, dialog boxes, and control bars displayed on the viewing screen. The root of all the window classes is *CWnd* (Figure 7.6). This class gives the common functionality to all windows like dragging and resizing characteristics. To draw graphics on an output device like a monitor screen MFC provides a device context abstraction in the class *CDC* and its derivatives (Figure 7.6). These classes provide a common interface for output that is independent of the type of device. By writing to the interface the output can be directed to either a monitor or a printer, for example.

Other classes in the MFC library include those for file services like *CFile* and *CArchive*, exception handling like *CException*, simple data values like *CString* and *CPoint*, data collections like *CArray* and *CList*, and database support like *CDatabase* and *CDaoDatabase* (Figure 7.6).

A majority of classes in the MFC library are derived directly or indirectly from class *CObject*. This virtual class provides support for a number of capabilities for its derived classes including debugging and file input/output. Generally, application software classes are also derived from the class *CObject* either directly or indirectly.

The MFC library is an abstraction of the Windows application programming interface (API). The Windows API is basically a non-OOP C-language interface that provides thousands of functions, messages, data structures, and data types needed to program in the Windows environment. MFC encapsulates the most common functionality of the API into an object-oriented interface and gives a more logical and conceptual view to the API. However, it still retains the power and flexibility of the underlying Windows API and makes it available to the software engineer. Development of a complex software system using MFC library requires some effort. This is because:

1. MFC is a framework of classes that interact with one another (understanding this interaction is very important as an application program has to "hook" into the framework),

2. MFC is rather large with about 200 classes, and

3. MFC provides an object-oriented encapsulation only for the most simple cases. In more complex situations, the software engineer has to develop an extension using the Windows API (Shepherd and Wingo, 1996).

7.8   AN APPLICATION ARCHITECTURE FOR THE CONSTRUCTION DOMAIN

Application development often requires the use of multiple specialized frameworks. In the design of complex software systems, it is essential that the role of each framework be defined accurately. Further, their dependencies must be clearly outlined to avoid any conflicts. In this work, a layered approach is used (Baumer *et al.*, 1997). This approach is based on high-level object decomposition and generalization of the application domain. This results in self-contained modules that can be implemented by one or more frameworks. Further, the *a priori* separation clearly outlines the scope of the frameworks and avoids duplication of functionality in their development. Note that a framework is a collection of cooperating classes relevant to a specific domain. Therefore, a related subset of a framework's classes may also be called a framework. For example, the MFC library is a framework but it can also be thought of as the collection of general-purpose application architecture, user-interface, and data management frameworks.

A high-level application architecture for the construction management domain is shown schematically in Figure 7.7. Three levels of abstractions are modeled. The outermost layer is the shell layer. All other layers depend on and use the services of this layer. Typically it provides commonly used data structures, mathematical functions, client/server middleware (low-level transaction management software), and request brokers (software that manages cooperation and communication among heterogeneous software components) for distributed computing. The shell layer is closely linked to the operating system, and its implementation in the form of frameworks is usually available for most application development environments.

Depending on the shell layer is the productivity layer, which is subdivided into database and user-interface layers. The database layer provides an interface to the applications for data management, storage, and retrieval. The frameworks in this layer may support OO or relational database management under either a centralized or distributed environment. The frameworks in the user-interface layer aid in the design of user-friendly application/user interactions. Depending on both the shell and productivity layers is the application

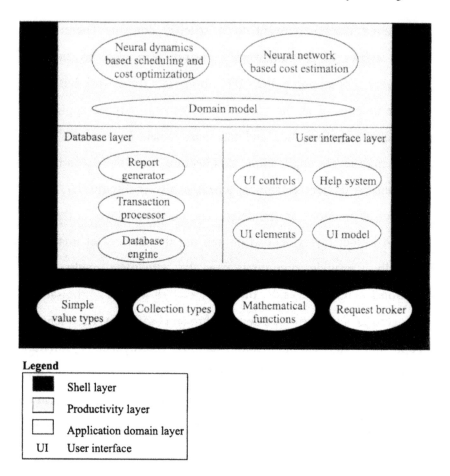

Figure 7.7   Schematic high-level view of the application architecture for the construction management domain

domain layer which contains domain specific details. Generally, this layer contains algorithms and models for the solution of problems in the domain. It is often subdivided to further categorize and generalize the application domain requirements. One or more frameworks may be used to implement this layer.

Figure 7.8 shows a detailed view of the architecture in the form of a package diagram. This diagram shows the breakdown of the application into packages and their dependencies. A package is a collection of related software elements. These elements may be classes, components, or frameworks. In Figure 7.8, the packages represent collection of classes. The dashed-line arrows indicate dependency of a package on another. A software dependency exists if any change in a package requires a change in the dependent package. Note that software dependencies are not transitive. That is, if package *A* depends on package *B* and package *B* depends on package *C*, package *A* may not necessarily depend on package *C* in terms of code implementation.

The application domain layer of the high-level architecture view is divided into two packages: a *Domain* package and an *Application* package (Figure 7.8). This is done because all application areas in construction scheduling, cost optimization, and management share common domain information. A framework-based implementation of this layer is presented in the next chapter.

The *Domain* package provides a common information base for software development in construction engineering. It contains organizational, construction schedule, and construction cost software building blocks that encapsulate domain specific knowledge and dependency logic in a reusable format. This layer depends on the *File and database support* package and the *Miscellaneous support classes* package of the MFC library.

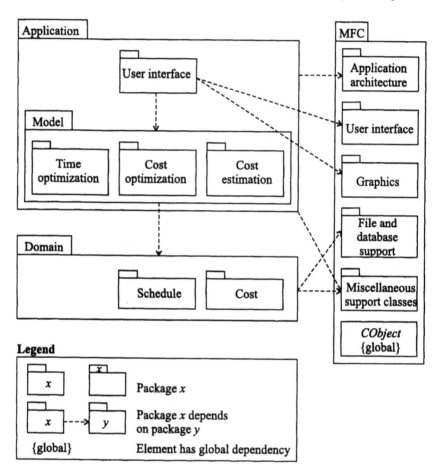

Figure 7.8   Package diagram for the construction domain application architecture

The *Application* package has two nested packages: *User interface* and *Model*. The package *Model* is a collection of various computational models and methods in the domain. This package depends on the *Domain* package and provides program logic and control for the solution of specific problems in the domain. The package *User interface* provides a unified user interface for the applications. This package depends strongly on the general-purpose *User interface* and *Graphics* packages provided by the MFC library.

The OO information model and ideas presented in this chapter have been implemented in a prototype software system for management of construction projects, called CONSCOM, which is described in the next two chapters. The MFC library is used for all general purpose functionality required for the development of CONSCOM. Software dependencies also exist between packages in the MFC library. However, these are not shown in Figure 7.8 for simplicity. By factoring out common concepts it is easier to maintain uniformity and consistency among applications in an area. The software design details of the domain-specific framework for CONSCOM is presented in the next chapter.

7.9   BRIEF DESCRIPTION OF CLASSES IN FIGURE 7.6

CArchive: Provides data streaming support for input and output of objects from a permanent storage.

CArray: Template-based array data structure

CCmdTarget: Provides user- and system-generated message handling support.

CDatabase: Provides basic database support.

CDC: Abstracts a device-context such as an output screen or a printer.

CDialog: Abstracts a dialog box display in the Windows environment.

CDocTemplate: Manages document and screen views in an application.

CFrameWnd: Abstracts a frame window display in the Windows environment.

CGdiObject: Base class for graphic output objects such as brushes and pens.

CList: Template-based list data structure (a collection in which each element maintains a pointer to its previous and next element).

CMap: Template-based map data structure (a collection in which each element is identified by a unique identification string).

CMenu: Abstracts a menu display in the Windows environment.

CObject: Provides support for debugging, diagnostics, dynamic object identification, and serialization (object storage and retrieval).

CPen: Abstracts a pen object (for drawing) for output purposes.

CRect: Encapsulates the coordinates of a rectangle.

CString: Provides support for managing strings (collection of characters).

CView: Manages the display of the data on the viewing screen.

CWinApp: Encapsulates the execution of a single-threaded
                  Windows application.

CWinThread: Encapsulates a single program thread (process in the
                  operating system) in the Windows environment.

CWnd: The base class for all window elements (provides
                  common display functionality).

8

THE CONSCOM FRAMEWORK

8.1   INTRODUCTION

Software development is often a major cost in the total cost of development and distribution of a new technology. Further, the development of software is incremental and evolutionary in nature with new requirements and features requested by users incorporated into it over time. This in turn requires that the software model be based on a reusable and extensible architecture that can be evolved over time.

In Chapter 7, we presented an object-oriented (OO) information model for construction scheduling, cost optimization, and change order management. The model represents a high-level software architecture for the development of extensible and compatible software systems in the construction management domain. This hierarchical or layered approach separates the key functionality of the system and allows for ease of development and maintenance especially since software components in a higher layer are independent of lower layer components. These software engineering techniques are fundamental in the development of reusable and extensible software systems.

In this and the following chapter we present a specific implementation of the model in a prototype software system called CONSCOM (CONstruction Scheduling, Cost Optimization, and Change Order Management). CONSCOM is implemented in Visual C++ using Microsoft Foundation Class library under the Windows environment. The software presently has over 19,000 lines of code with over 100 classes. This chapter presents the implementation details of the CONSCOM object-oriented model and application framework based on the application architecture developed in Chapter 7.

8.2   THE CONSCOM FRAMEWORK

8.2.1 Introduction

The CONSCOM framework is a white-box black-box framework. A framework is called a white box if it provides abstract classes only and no concrete classes. Use of such a framework requires the development and implementation of concrete classes that are derived from the framework classes. This in turn requires an understanding of the framework design. Hence, the name white box. On the other hand, black box frameworks provide default implementations that can be used directly with little or no change. The CONSCOM framework has elements of both. It defines an interface that can be implemented as desired. It also provides default implementation of the new problem solving techniques for scheduling, cost optimization, and cost estimation (Adeli and Karim, 1997b; Adeli and Wu, 1998).

To facilitate reuse and sharing of the framework in the Windows environment the framework is packaged in a Windows dynamic link library (DLL). A DLL is a file that contains precompiled code (such as C++ classes) that can be used and shared at run-time without recompilation with each application. All applications that use the services of a DLL can link with the file at run-time. Further, if several applications on a computer use the same DLL only one copy is loaded into the memory and shared among the applications. DLLs provide a binary platform dependent technique for sharing and reusing code. A consequence of using precompiled code is ease of maintaining a standard and compatible version for use. In the case of the CONSCOM framework a common standard information model for construction engineering can be maintained and distributed.

The CONSCOM framework is built using software patterns. This makes the software design robust. Software patterns will also serve as documentation for the framework (Odenthal and Quibeldey-Cirkel, 1997). However, most of the software design patterns discussed and presented in the literature deal with generic situations as opposed to domain-specific situations. Generic software patterns are usually low-level software design ideas for creating, organizing, and maintaining objects. Domain-specific patterns, on the other hand, are usually more elaborate and provide domain-modeling ideas. In the development of CONSCOM generic patterns are used when applicable. The OO design principles are also exploited for cases where the documented patterns do not provide a solution. The software design concepts presented in this book for modeling construction management problems can be generalized into domain-specific software design patterns for construction management. In the following discussion software design pattern names are identified in small caps.

8.2.2   Object Model

Figures 8.1–8.9 show the key classes in the CONSCOM framework, their characteristics (abstract or concrete), and their object models. The notation in these and all subsequent figures are based on the new standard Unified Modeling Language (UML) (Fowler and Scott, 1997). Brief description of classes in Figures 8.1–8.9 are given at the end of this chapter. Classes starting with the prefix 'C' are MFC classes while those starting with 'CI' are implementation classes for construction scheduling, cost optimization, and cost estimation.

The classes shown in Figures 8.1 and 8.2 correspond to the *Domain* and *Application* packages, respectively, described in Chapter 7. The *Domain* package (collection of classes) encapsulates the construction scheduling and cost knowledge. This package is the computational engine of the software system. The *Application* package provides support for application-specific domain knowledge and the application's user interface. Note that Figures 8.1 and 8.2 do

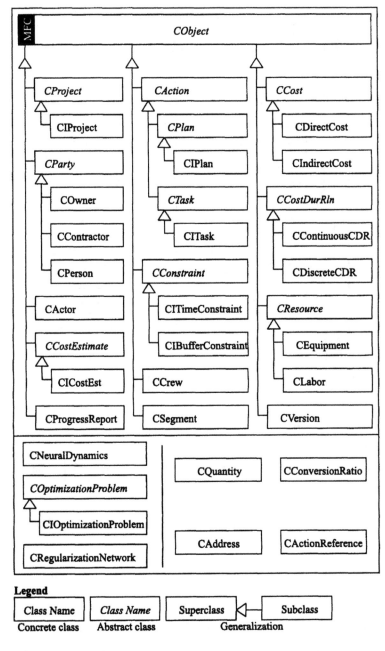

Figure 8.1   Class diagram of the *Domain* package in CONSCOM

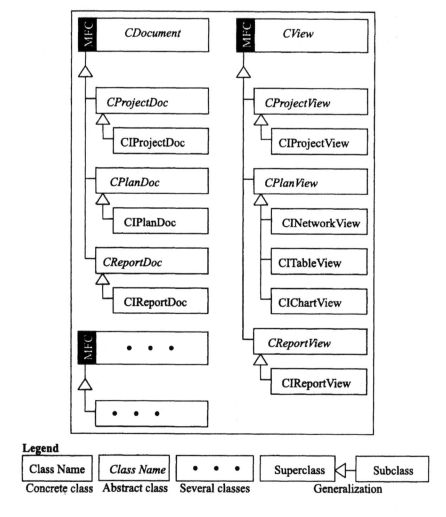

Figure 8.2   Class diagram of the *Application* package in CONSCOM

not show all the classes that support the user interface of CONSCOM.

Most of the classes have the MFC class *CObject* as the root to take advantage of the services it provides for object storage and retrieval. Note that most of the higher level classes are abstract. This is a fundamental design concept in frameworks that allows customization through subclassing. The basic framework (consisting of the abstract classes only) just provides an interface. Derived classes bind the interface to a specific implementation. An abstract class controls which implementation class is instantiated.

8.2.3   Model Description

The focal point in construction engineering and management is the construction project. A project (construction or otherwise) is defined for every collection of related activities that needs to be completed under certain constraints and hence needs to be managed. Some key elements of a construction project are the proposed and implemented plan of work, the cost and time reference, the cost estimate, and the descriptive progress report. Effective management requires that a track record of all these information elements be maintained. Figure 8.3 shows a class diagram of how this is modeled in CONSCOM. Attributes and operations shown in Figures 8.3–8.9 are defined at the end of this chapter. The classes *CProject*, *CPlan*, *CCostEstimate*, and *CProgressReport* abstract their real-world equivalents. Objects of each one of these classes are associated with a *CVersion* object. Class *CVersion* maintains the current version number, the time of creation and last update, and a pointer to the person or organization that made the last change. The class *CActor* abstracts a role or job

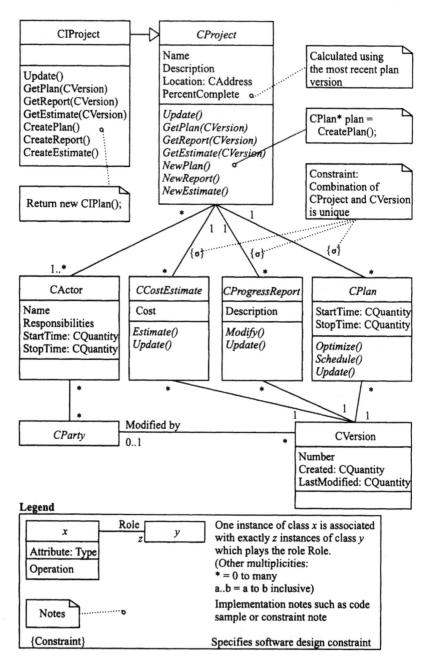

Figure 8.3   Construction project model in CONSCOM (Class
            ***CProject*)**

function. A project has several job functions associated with it such as manager, scheduler, and supervisor. The class *CProject* plays the role of an abstract factory as described in the software pattern ABSTRACT FACTORY (Gamma *et al.*, 1995). This software pattern describes how to solve the problem of creating several related objects without explicitly specifying their concrete classes. For example, class *CProject* only defines the interface (*NewPlan()*) for creating a new plan (Figure 8.3). The subtype of *CPlan* object created will depend on which concrete class of *CProject* is used. In Figure 8.3, *CIProject* creates objects of class *CIPlan*. Therefore, the set of objects created will depend on which 'factory' is used.

The method or operation *NewPlan()* (and other new operations) in class *CProject* represents a factory method as defined in the pattern FACTORY METHOD (Gamma *et al.*, 1995). This pattern describes a technique to delegate the creation of an object to its concrete subclasses. This technique uses the dynamic binding construct in OOP languages in order to make it transparent to the user. The classes *CProject*, *CPlan*, *CTask*, *CConstraint*, *CCostEstimate*, *CProgressReport*, and *COptimizationProblem* all provide a factory method to isolate the object creation process from the user. The type of the object created will depend on the class to which the pointer points. These classes are also the hot spots of the CONSCOM framework. A hot spot represents a point of variability, which can be used to customize the framework (Schmid, 1996, 1997).

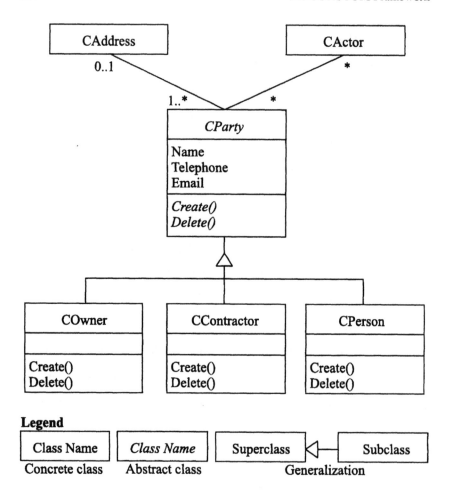

Figure 8.4   Organization model in CONSCOM (Class *CParty*)

Many individuals and organizations become associated with a construction project throughout its life span. However, in several situations no difference is made between whether the associated party is an individual or an organization. For example, the role of the project's owner may be played by either an individual or an organization. Figure 8.4 shows how this concept is modeled in CONSCOM. This software design is based on the PARTY pattern (Fowler, 1997). The class *CParty* abstracts characteristics common to persons and organizations such as contact address and job function or role.

Effective representation of measurements is essential in any software system. The construction engineering domain has a wide range of measurement types such as distance, time, and volume. Further, a wide range of measurement units are used such as hectares or square meters for measuring areas. To maintain the integrity of the measurement and the calculations based on them both the value and the unit of the measurement must be encapsulated in a single object. The class *CQuantity* (Figure 8.5a) provides this functionality. It also allows conversion of a value from one unit to another. This conversion is done automatically whenever any arithmetic or logical operators are used on objects of *CQuantity*.

In the CONSCOM framework all measurements are of type *CQuantity*. Ratios for converting a value from one unit to another are maintained by an object of class *CConversionRatio* (Figure 8.5b). Only one instance of *CConversionRatio* is required. Multiple instances are not only unnecessary but also create the problem of data consistency maintenance among all the instances. To prevent multiple instances of class *CConversionRatio*, it is designed as a singleton as defined in the software pattern SINGLETON (Gamma *et al.*, 1995).

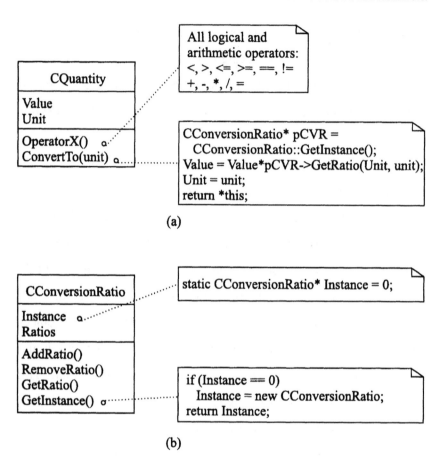

**Figure 8.5   Measurement model in CONSCOM: (a) Class *CQuantity*;
(b) Class *CConversionRatio***

The plan is the most significant element in the management of a construction project. Traditionally, a plan is defined as the timetable of the tasks that needs to be carried out. In other words, a plan tells us when each task will be executed in the future. In this work a broader view of the construction plan is adopted. The construction plan is considered to represent the current state of the project. This includes the tasks that have been completed, those that are in progress, and those proposed to be carried out in the future. Similarly a construction task may be a proposed task, a completed task, or an in-progress task. Using these broader conceptual views it is possible to capture the state of work at any given point in time. This is very important in change order management and conflict resolution because comparisons can be made much more easily. Note that unless a plan has been implemented it can be scheduled and optimized no matter which state it is in.

Figure 8.6 shows how the construction plan and task are modeled in CONSCOM. The classes *CPlan* and *CTask* are defined as specialization (subtypes) of the class *CAction*. The class *CAction* abstracts the common characteristics in classes *CPlan* and *CTask* and gives them similar interfaces. For example, both construction plans and tasks have a time reference and both of them can be scheduled. To model this in software a supertype is created which abstracts the common characteristics. A status variable identifies which state a plan or task is in. The states relevant to construction management are: Proposed, started, completed on time, delayed, abandoned, and suspended. Each object of class *CAction* is associated with a previous version of itself. This generates a hierarchy that captures the history of changes that have occurred over time. Each object of class *CAction* ensures that the previous version object pointed to is valid.

Construction task and plan have some similarities of behavior as described above. A construction plan, however, is a collection of

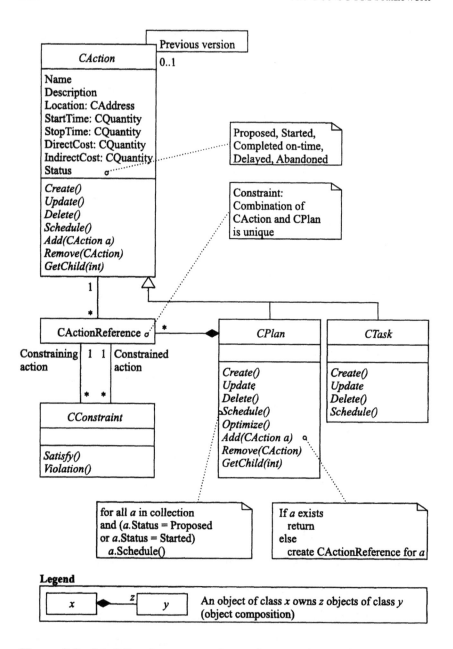

Figure 8.6 Model of construction plan, task, and scheduling constraint in CONSCOM

construction tasks. In a computer model this structural difference must be hidden from the user when executing a common behavior operation. Figure 8.6 shows how this is modeled in CONSCOM. The class *CPlan* composes objects of class *CActionReference* that holds a reference to a single object of class *CAction*. The class *CAction* provides an interface for managing collection of objects. These operations are only implemented in the composite class (*CPlan*) and not in the child class (*CTask*). This design is based on the COMPOSITE pattern (Gamma *et al.*, 1995). The composition relationship between *CPlan* and *CActionReference* means that an object of class *CPlan* owns objects of class *CActionReference*. In other words, the creation and destruction of *CActionReference* objects depend on the *CPlan* object. Creation of a *CActionReference* object by an object of *CPlan* [in Add(*CAction* a)] will occur only when the *CAction* object being added is not already present. This prevents a *CPlan* object from having multiple references to the same *CAction* object. Note that with the present software design it is also possible to model a plan that contains a combination of both tasks and plans.

Scheduling constraints between *CAction* objects (tasks and/or plans) are modeled by the *CActionReference* and *CConstraint* classes (Figure 8.6). The class *CConstraint* encapsulates a single scheduling constraint between two objects of class *CAction*. It provides an interface for calculating constraint violation and determining constraint satisfaction. In CONSCOM, two concrete classes are provided which implement these operations. The class *CITimeConstraint* models the precedence relation constraint, while the class *CIBufferConstraint* models time and distance buffer constraints. These implementations are developed based on the general scheduling model described in Chapters 5 and 6.

In construction scheduling, accurate modeling of the construction task is essential. A task may be repetitive or non-repetitive and, if it is repetitive, it may be executed by a single crew or multiple crews, and/or it may require distance modeling. This last concept is essential in linear construction projects, such as highway construction. Also, effective cost control requires that the resources required and used by a construction task be accurately modeled. A mathematical model of these concepts was presented in Chapter 5. Its representation in CONSCOM is shown in Figure 8.7. The class *CCrew* represents a construction crew. The class *CSegment* encapsulates a segment or location of work for a crew. An object of class *CSegment* also encapsulates the job conditions and quantity of work for that segment. Each *CCrew* object owns one or more *CSegment* objects. However, two *CSegment* objects cannot have overlapping locations. This constraint is enforced by the *CCrew* object that is creating the *CSegment* object. The class *CResource* abstracts a construction resource such as construction equipment and labor. Depending on the productivity data for the resource a cost duration relationship can be developed. This information is

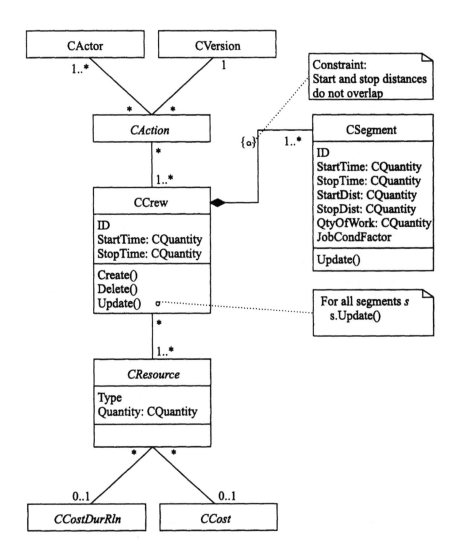

Figure 8.7   Model of construction task in CONSCOM

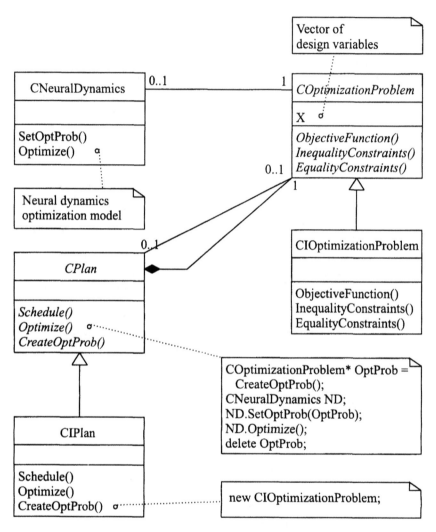

Figure 8.8   Construction cost optimization model in CONSCOM

encapsulated by the class *CCostDurationRln*. The class *CCost* provides an interface for managing the construction cost.

Decisions involving change order and conflict resolution often require cost comparisons among different versions and states of the construction project. The scheduling/cost optimization model presented in Chapters 5 and 6 provides the basis for consistently reliable evaluation of the cost of a project at any point in time. Figure 8.8 shows how the general neural dynamics model is used to solve the problem of construction cost minimization. The class *CNeuralDynamics* encapsulates the neural dynamics optimization algorithm. The class *COptimizationProblem* defines the interface required for an optimization problem in order to be solved by the neural dynamics model.

In CONSCOM, the construction direct cost optimization problem is defined by an object of type *CPlan*. The goal is to map this problem (defined by an object of type *CPlan*) to another object that has an interface of class *COptimizationProblem*. In this design two software patterns are used. The ADAPTER (Gamma *et al.*, 1995) pattern solves the problem of how to adapt one interface to another and the TEMPLATE METHOD pattern solves the problem of how to choose and create the correct adapter object. The Optimize() operation in *CPlan* is the template method. The type of *COptimizationProblem* created depends on the implementation of the *CreateOptProb*() operation. In the present case, the *CIPlan* object creates an object of *CIOptimizationProblem*.

Figure 8.9 shows an example of how an MFC application document is associated with the corresponding domain information. For example, class *CProjectDoc* encapsulates an application document that contains construction project information. The

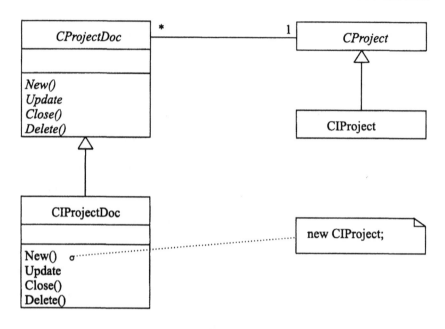

Figure 8.9   Model of application document in CONSCOM

FACTORY METHOD pattern is again used to control which concrete subtype of *CProject* is created.

A major concern of the construction project owner is the effective management of change orders. A change order can be initiated by the contractor, the owner, the designer, or any other stakeholder in the project. As a result of a change the contractor may claim an extension in the project duration in addition to compensation for any additional work required. To support its claim an updated schedule is submitted to the owner for review. The owner's change order management process can be divided into two steps. First, a qualitative review of the schedule is carried out to identify any omissions, errors, inaccuracies, and non-compliance with the contract documents. Second, a quantitative analysis is done to determine the impact of the change and its comparison with what is being claimed by the contractor. Contractors often modify the schedule to exaggerate their claims. Therefore, these steps must be executed with care and accuracy. Further, consistency and reliability of the owner's decision is essential for it to be acceptable to the contractor. The CONSCOM framework provides all these capabilities to automate and expedite the decision of the project owner when faced with a change order.

The use case diagram for construction change order management is shown in Figure 8.10. A use case diagram captures the users and their uses of a software system. A use case represents a requirement of a software system that must be satisfied. When faced with a change order the owner's primary use of the software system is decision support. This is represented by the *Change order management* use case (Figure 8.10). The qualitative and the quantitative steps of the change order management process are represented by the use cases *Review plan* and *Analyze scenarios*, respectively. The *Review plan* use case involves a compliance check with the contract documents and a comparison with a previous version of the plan. If any error or non-compliance is identified the contractor is asked to send an updated plan.

The *Analyze scenarios* use case involves the generation of several construction plan scenarios and carrying out time-cost trade-off analyses on them. Typical scenarios include trying different

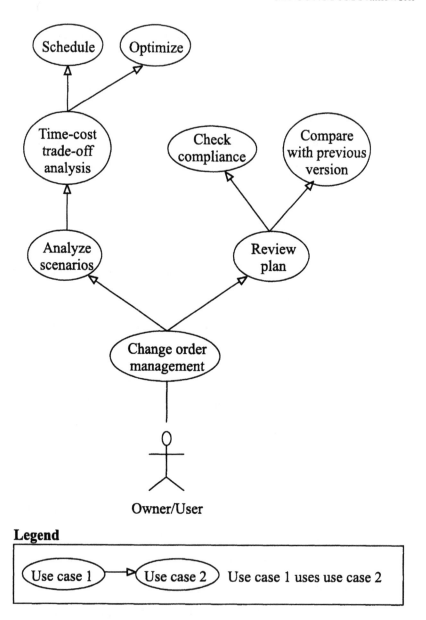

Figure 8.10 Use case diagram of construction change order management

project durations or following an alternate logic for the construction tasks. These analyses are based on scheduling and cost optimization of the generated scenarios. Therefore, the results are consistent across all scenarios. From these analyses the owner is able to make a decision on the reasonableness of the claims made by the contractor. Note that the owner can also analyze different construction plan scenarios to determine the feasibility of a change that it (or he) wants in the project.

Figure 8.11 shows an object interaction diagram in CONSCOM to solve the problem of change order management. The sequence of operations executed by each object is identified from top to bottom along the dashed line. The latest plan is analyzed under multiple scenarios to determine the exact impact of the change. The process terminates with updating of the *CProject* object and recording of the decision taken in the project progress report.

8.3   CONCLUSION

In this chapter we presented the CONSCOM object model and framework as a specific implementation of the application architecture for construction management software systems presented in Chapter 7. Key software design concepts and ideas necessary for the development of extensible, compatible, and maintainable software systems and used in CONSCOM were highlighted. CONSCOM is a prototype software system for construction scheduling, cost optimization and change order

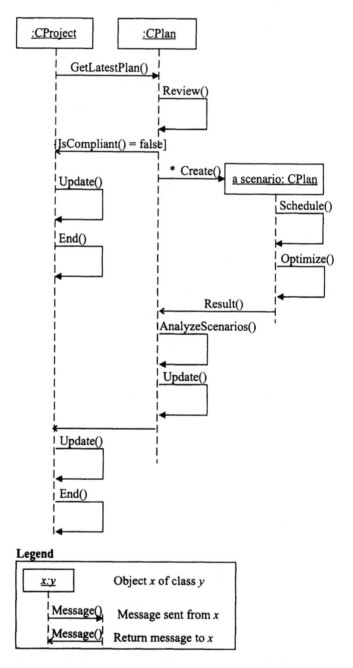

Figure 8.11 Object interaction diagram showing use of CONSCOM for construction change order management

management that can be used by both owners and contractors for the effective management of construction projects. CONSCOM is described in further detail with illustrative examples in the next chapter.

8.4   BRIEF DESCRIPTION OF CLASSES IN THE CONSCOM FRAMEWORK (FIGURES 8.1–8.9)

CAction: Provides an interface for managing actions (tasks and plans).

CActionReference: Encapsulates a reference to a *CAction* object.

CActor: Encapsulates a role played by a person or organization (participant).

CAddress: Encapsulates a street address.

CConstraint: Encapsulates a construction scheduling constraint.

CContinuousCDR: Encapsulates a continuous cost duration relation.

CContractor: Abstracts a contractor of a construction project.

CConversionRatio: Encapsulates ratios for the conversion of values from one unit to another.

CCost: Provides an interface for managing construction costs.

CCostDurRln: Encapsulates a cost duration relation.

CCostEstimate:	Provides an interface for managing construction cost estimates.
CCrew:	Abstracts a construction crew.
CDirectCost:	Provides support for construction direct cost management.
CDiscreteCDR:	Encapsulates a discrete cost duration relation.
CEquipment:	Abstracts a construction equipment.
CI:	Class names starting with 'CI' are implementation classes for the corresponding class name starting with 'C' in CONSCOM.
CIBufferConstraint:	Provides support for distance or buffer constraints.
CIndirectCost:	Provides support for construction indirect cost management.
CITimeConstraint:	Provides support for time constraints.
CLabor:	Abstracts information of construction labor.
CNeuralDynamics:	Abstracts the neural dynamics model for the solution of optimization problems.
COptimizationProblem:	Provides an interface for the solution of optimization problems using the neural dynamics model.
COwner:	Abstracts an owner of a construction project.

CParty:	Base class of *CContractor*, *COwner*, and *CPerson* (encapsulates their common features).
CPerson:	Abstracts a person.
CPlan:	Provides an interface for managing construction plans.
CPlanDoc:	Provides an interface for a construction plan application document.
CPlanView:	Provides an interface for a construction plan application view.
CProgressReport:	Provides support for managing progress reports.
CProject:	Provides an interface for managing construction projects.
CProjectDoc:	Provides an interface for a construction project application document.
CProjectView:	Provides an interface for a construction project application view.
CQuantity:	Encapsulates a measurement value and its unit.
CRegularizationNetwork:	Abstracts the regularization neural network model for cost estimation.
CReportDoc:	Provides an interface for a construction progress report application document.

CReportView:	Provides an interface for a construction progress report application view.
CResource:	Abstracts a construction resource.
CSegment:	Abstracts a segment of construction work.
CTask:	Provides an interface for managing construction tasks.
CVersion:	Provides support for version control.

8.5   BRIEF DESCRIPTION OF THE ATTRIBUTES AND OPERATIONS SHOWN IN FIGURES 8.3–8.9

8.5.1   Attributes

Description:	Description of an object.
DirectCost:	Direct cost.
Duration:	Duration of an activity.
Email:	Email address.
ID:	Identification.
IndirectCost:	Indirect cost.
Instance:	Instance of an object.
JobCondFactor:	Job condition factor.
Location:	Address information.
Name:	Name of an object.
PercentComplete:	Percentage of a work that is complete.
QtyOfWork:	Quantity of work required.
Quantity:	Quantity.
Ratios:	Conversion ratios between units.

Responsibilities:	List of responsibilities.
StartDist:	Starting distance of a segment of work.
StartTime:	Starting time of an action.
Status:	Status information.
StopDist:	Stopping distance of a segment of work.
StopTime:	End time of an action.
Telephone:	Telephone number.
Type:	Type information.
Unit:	Measurement unit.
Value:	Numeric value.
X:	Vector of variables.

8.5.2 Operations

Add():	Adds an object to a collection.
AddRatio():	Adds a ratio to a collection.
ConvertTo():	Converts a measurement from one unit to another.
Create():	Creates an object.
CreateOptProb():	Creates an object of type *COptimizationProblem*.
Delete():	Deletes an object.
EqualityConstraint():	Evaluates the equality constraints of an optimization problem.
Estimate():	Estimates the cost of a project.
GetEstimate():	Returns a pointer to an object of type *CCostEstimate*.

GetInstance():	Returns an instance of an object.
GetPlan():	Returns a pointer to an object of type *CPlan*.
GetRatio():	Returns a ratio from a collection.
GetReport():	Returns a pointer to an object of type *CProgressReport*.
InequalityConstraint():	Evaluates the inequality constraints of an optimization problem.
New():	Creates a new object.
NewEstimate():	Creates a new object of type *CCostEstimate*.
NewPlan():	Creates a new object of type *CPlan*.
NewReport():	Creates a new object of type *CProgressReport*.
ObjectiveFunction():	Returns a value of the objective function of the optimization problem.
OperatorX():	Implements arithmetic and logical operators of type *CQuantity*.
Optimize():	Minimizes the cost of a construction plan.
RemoveRatio():	Removes a ratio from a collection.
Satisfy():	Satisfies a scheduling constraint.
Schedule():	Schedules a construction plan or task.
SetOptProb():	Sets up a reference to an optimization problem.
Update():	Updates the state of an object.

Violation(): Returns the amount of violation of a scheduling constraint.

A NEW GENERATION SOFTWARE FOR CONSTRUCTION SCHEDULING AND MANAGEMENT

9.1   INTRODUCTION

The CONSCOM object model and framework was presented in Chapter 8. In this chapter we describe novel features of CONSCOM and its user-interface with particular emphasis on the integration of the modeling, control, and management features. The use of these features is subsequently illustrated by an example project.

9.2   INTEGRATED CONSTRUCTION SCHEDULING AND COST MANAGEMENT

As stated in Chapter 1, despite its shortcomings the CPM is still the most widely used method in practice. None of the other methods presented over the years has gained widespread acceptability. Three reasons can be cited for this lack of acceptance of newer techniques. First, there is the compatibility problem. To ease migration to the new technique it must provide all the modeling capabilities of the existing technique (CPM). If the new technique is a superset of CPM then the initial training and learning cost is minimized as the user can learn the new capabilities gradually with time. Second, the techniques presented over the last two decades do not provide substantial improvements or advantages to justify their widespread use by the construction industry. Third, there is the issue of ease of use and software maintenance.

To overcome the first two problems, in Chapters 5 and 6 we developed a new integrated construction scheduling and cost management model. This model was initially motivated by the need to handle repetitive task projects such as highway construction. However, the model is general in applicability providing a complete set of features, including the four precedence constraint relationships that are considered standard in the CPM. In addition, the new model includes location modeling of tasks with time and distance buffer constraints, resource allocation features such as multiple crew assignment and management, resource allocation that can vary either linearly or nonlinearly with the duration of work, and construction progress tracking, control, and management. The mathematical model incorporates a robust cost minimization algorithm based on the recently patented neural dynamics model of Adeli and Park. This latter feature provides an essential tool to the user for time-cost trade-off analysis and change order management.

Although CONSCOM is based on a mathematically rigorous foundation, the scheduling concepts and structures used in this model are generalizations of those in CPM. Therefore, the users of CPM-based software systems can adapt to the new system without significant investment in learning new concepts. However, the solid mathematical foundation makes the implementation unambiguous and serves as a baseline for extension and further development.

Advanced object technologies were used in the development of CONSCOM. In doing so our motivation was to produce a prototype software system that is maintainable and extensible. The system's underlying software architecture, object model, and framework were described in Chapters 7 and 8. An advanced software model, however, would be of little use if the software system is not easy to use. To overcome this problem, we developed an integrated graphical user interface for construction scheduling, cost optimization, and change management that makes available all the features of the model in an easy to use and understand interface. These features are described and illustrated in this chapter.

9.3   FEATURES OF CONSCOM

CONSCOM is an advanced new generation software system for construction scheduling and management that includes a powerful scheduling model, has robust optimization capabilities, and provides strong change management features, all integrated into a compact

Integrated Management Environment (IME). Some of the key features of CONSCOM are delineated in the following paragraphs:

- CONSCOM features an advanced construction scheduling model. This model is a superset of all currently available models such as CPM plus new features such as:
  - Integrated construction scheduling and minimum cost model.
  - Support for a hierarchical work breakdown structure with tasks, crews, and segments of work.
  - Capability to handle multiple-crew strategies.
  - Support for location (distance) modeling of work breakdown structures (very useful for modeling linear projects such as highway construction).
  - A mechanism to handle varying job conditions.
  - Nonlinear and piecewise linear cost modeling capability for work crews.
  - Capability to handle time and distance buffer constraints in addition to all the standard precedence relationships.
  - Ability to provide construction plan milestone tracking.
- CONSCOM's computational engine is based on the recently patented robust and powerful neural dynamics optimization model of Adeli and Park. The neural dynamics model of Adeli and Park provides reliable cost minimization of the construction plan, time-cost trade-off analyses, and change order management.

- CONSCOM provides an integrated user-interface with all the tools, capabilities, and information necessary for effective control and management of construction projects.

- CONSCOM provides a context-sensitive help facility readily available at any point of execution of the software.

9.4   INTEGRATED MANAGEMENT ENVIRONMENT

CONSCOM provides an integrated user interface for effective construction management and control. The interface is an integrated management environment (IME) providing ready access to all the tools and information needed to plan, monitor, analyze, and control construction projects. Figure 9.1 shows a screen shot of CONSCOM's main window. Three types of output display windows are used to provide information to the user. The primary output display is the Plan View window (the two windows on the top-right in Figure 9.1). Each of these windows displays the details for a specific plan. All the tasks are listed together with their start time, finish time, duration, and cost. The two-part icon in front of each row indicates the status of the task. Two different colors are used for the two parts of the icon to give visual information as to whether the task is a proposed task or an implemented task (a task currently in progress).

Multiple plans may be opened in CONSCOM at a single time. All the plans that are open plus all the plans that were previously open in the current CONSCOM session are listed in the output

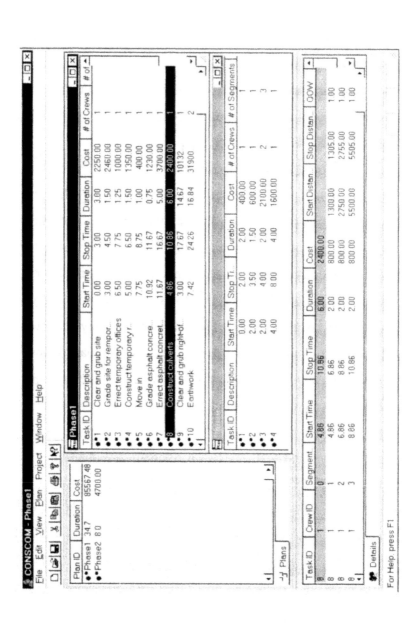

Figure 9.1 CONSCOM's main window

display called the Project Workspace window (left window in Figure 9.1). This output display lists all the plans in the workspace with their current cost and duration values. The Project Workspace window is the project management and control interface. The information available from this window facilitates the owner or the contractor to keep track of all the plans in the project. Further, if CONSCOM is used for change order management then this window allows one step access to all the versions of the plan enabling quick and effective decision making. Similar to the Plan View windows the two-part icon in front of each plan in the Project Workspace window indicates the status of the plan.

The third output display is the Task Details window (the bottom window in Figure 9.1). This window shows the detailed information for each work breakdown structure of the selected task. The information in this window is especially useful for multiple-crew and multiple-segment tasks. The information displayed includes the start time, finish time, duration, and cost for the work of each crew and each segment of the work of each crew. In addition, for each segment of the work, the start and stop locations, the quantity of work, and the job condition factor are also displayed. Further information about each task that is less frequently required is provided in dialog boxes (an input/output window that is usually not resizable). Figure 9.2 shows the Modify Task property dialog that allows modification and display access to all the information for a selected task. Certain

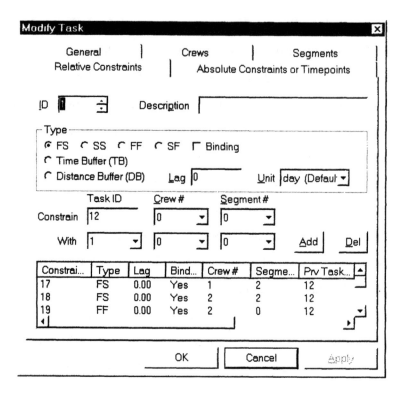

Figure 9.2   Task information access and modification dialog box

information such as the constraints, cost-duration relationship, and descriptions of crews is available from this property dialog only.

9.5   USER INTERFACE CHARACTERISTICS

The highway construction project example presented in Section 6.4 is used to demonstrate the modeling capabilities of CONSCOM. This example uses some of the new scheduling features provided by the model. This plan cannot be modeled by CPM or CPM-like networks currently available in commercial packages. In this section, we describe the handling of these features by CONSCOM's user interface.

The work for the construction of the 2-lane 5-km long highway is divided into 7 repetitive and 7 non-repetitive tasks (Table 6.1). This project requires the following modeling features: multiple crews, multiple segments of work per crew, location modeling of work, distance and time buffer constraints, work continuity constraints, and job condition factors. Both linear and nonlinear relationships are used to describe the direct cost-duration relationship of crews. A minimum-duration plan generated by CONSCOM is shown in Figure 9.3. Note that in CONSCOM no distinction is made between a repetitive and a non-repetitive task. The differentiation is only for user convenience. Computationally, both types of tasks are handled similarly by CONSCOM.

To illustrate the data input process and show some of the modeling capabilities of CONSCOM consider adding a new task 10 to the plan. In CONSCOM, entry and modification of data for a task is handled by four dialog boxes. When adding a new task these dialog boxes are presented in a logical sequence. When the user is modifying a selected task all the dialog boxes are presented simultaneously. The four dialog boxes showing the input data required for task 10 are shown in Figures 9.4 to 9.7. General information of the task is entered in the General dialog box (Figure 9.4). In this figure, ID is an alphanumeric string that uniquely

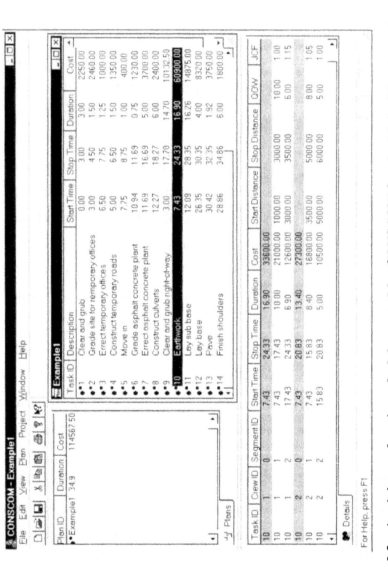

Figure 9.3 A minimum duration schedule created by CONSCOM for the 2-lane 5-km long highway construction project

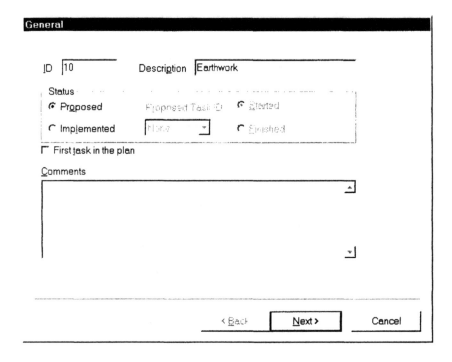

Figure 9.4   General information dialog box (Input for task 10)

identifies the task in the plan. The description and comments are optional fields. In this example, the task is a proposed task.

However, if progress of work is being monitored or change order scenarios are studied a corresponding implemented task can be added by selecting the appropriate button. An implemented task may or may not have a corresponding proposed task. For example, the contractor may add and start working on a new task because of an unexpected working condition on the field which was not anticipated in the original proposed plan. This will require a new implemented task that has no corresponding task in the proposed plan. For further

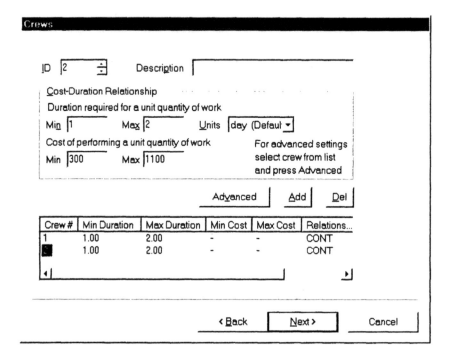

Figure 9.5   Crews dialog box (Input for task 10)

classification of the status of an implemented task the task can be set as either one that is in progress or one that is finished. When a task has the implemented status its start time and cost are fixed and it will not take part in scheduling and cost optimization.

The work crew information for the task is entered in the Crews dialog box (Figure 9.5). Each task can have multiple crews that are uniquely identified by a numeric value. Associated with each crew is a direct cost-duration relationship. The default relationship is a two-point linear relationship corresponding to the minimum cost (maximum duration) and maximum cost (minimum duration) data

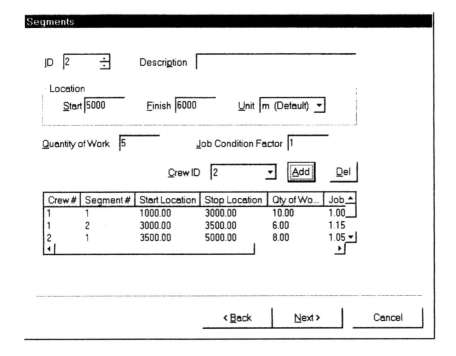

Figure 9.6   Segments dialog box

points. If a piecewise linear or a nonlinear relationship is desired then it can be specified in the cost-duration relationship dialog box (Figure 9.8). The piecewise linear relationship is defined by a finite number of cost-duration data points which are then connected by straight lines.

The nonlinear relationship is modeled in the following form:

$$C(d) = \frac{f(d)}{g(d)}$$

where $C(d)$ is the cost of completing the work in duration d, and $f(.)$, $g(.)$ are the numerator and the denominator fourth order polynomial

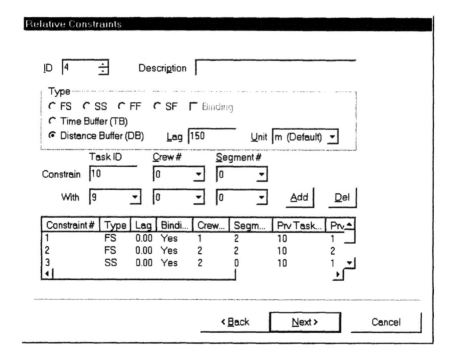

Figure 9.7   Relative Constraint dialog box

expressions, respectively. In the present example, task 10 has two crews and both of them have the identical nonlinear relationship $C(d) = (1600 + d)/d$ where $C(d)$ is the cost for completing work in duration d. The cost-duration relationship is bounded by the minimum and maximum duration values that have to be specified.

The third input dialog box, called Segments dialog box (Figure 9.6), gathers information for the segments of work for each crew. Each segment is uniquely identified for each crew by a numeric value. The breakdown of work into segments is necessary to capture the changed conditions at each stage of the crew's work. These

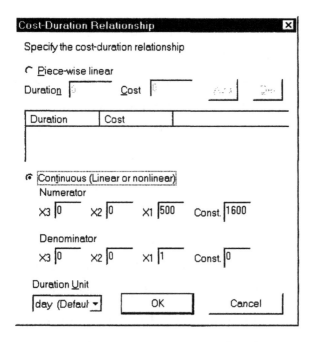

Figure 9.8   Cost-duration relationship dialog box

variations may include the difference in location, change in quantity of work, or a better or worse work condition than originally expected. In CONSCOM, locations are modeled using distances. This is useful in linear projects such as highway construction where it is necessary to track the location of work not only for progress monitoring purposes but also to ensure that sufficient distances are maintained between tasks. Distance buffers are required when, for example, sufficient space is required for equipment and labor to perform optimally and safely. When distance modeling is not needed the location fields should be set to zero. This does not mean that location is not considered in modeling but locations of work are not defined by linear distance.

Task 10 has two crews and two segments of work per crew (Figures 9.5 and 9.6). Crew 1 works over the distance 1000 to 3500 m while crew 2 works over the distance 3500 to 6000 m. The work of crew 1 is broken down into two segments from 1000 to 3000 m and from 3000 to 3500 m (Figure 9.6). This breakdown is done to model the more difficult job conditions in the 3000–3500 m section. This segment is assigned a job condition factor of 1.15 that reflects a 15 percent increase in time needed to move a unit quantity of earth as compared to that required in the first segment of work. Another reason to break down the work of a crew is when the quantity of work required per unit length of the highway is not constant. Segments of work are selected so that the quantity of work per unit length in each segment remains roughly constant.

The last input dialog box, called the Relative Constraint dialog box (Figure 9.7), is used to specify constraints on the task. CONSCOM supports all the standard precedence constraints. In addition, it also supports time and distance buffer constraints. A constraint can be specified on any segment of the task, on any crew of the task, or on the whole task. Similarly, the constraining element can be any segment, crew, or task. This flexibility in specifying constraints allows multiple-crew strategies, work continuity, and other resource-based constraints to be modeled effectively. Each constraint can also have a time or distance lag value. The two crews of task 10 have a start-to-start relationship. With this constraint both crews will start work at the same time. Also, note that this constraint is specified as a binding constraint. This means the constraint is an equality constraint. On the other hand, a non-binding constraint is an inequality constraint that is satisfied as long as its value is less than or equal to zero. For example, the 150 m distance buffer constraint between task 10 and task 9 is non-binding. This ensures a minimum distance of 150 m (not exactly 150 m) is maintained between the work crews of task 10 and task 9. The two segments of work of each crew have a work continuity constraint to ensure continuity of work.

Data in CONSCOM can be entered in any appropriate unit. The user can define a default set of units for each plan or project. These are the units in which all output is displayed. However, at all the data entry points a set of appropriate units is available to the user to choose from (see Figures 9.4–9.7). CONSCOM will convert the entered data from the specified unit to the default unit automatically.

In CONSCOM a plan can be scheduled in three ways. First, the plan may be scheduled so that all tasks are completed with minimum duration. Second, the plan may be scheduled so that all tasks are completed with minimum cost. Third, the plan may be scheduled for given fixed durations of tasks. The example schedule shown in Figure 9.3 is for minimum duration.

9.6   EXAMPLE – RETAINING WALL PROJECT

In this section, an example is presented that illustrates the use of CONSCOM for construction cost minimization and change order management.

The project is a 6 m high and 30 m long cast in-place reinforced concrete retaining wall (Figure 9.9). These structures are commonly used alongside entry and exit ramps on highways. The work required for the project is divided into five tasks (Table 9.1). Each task has one work crew that performs consecutively on 10 m lengths of the wall. That is, each crew has three work segments or locations consisting of a 10 m length of the wall. The direct cost-duration relationship for each crew is a straight line defined by minimum and maximum cost/duration data points (Table 9.1). The work

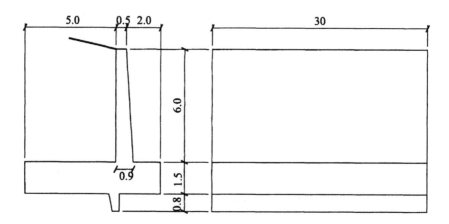

All dimensions are in meters

Figure 9.9   The retaining wall construction project

Table 9.1   Task ID, description, and direct cost duration relationship for each task of the retaining wall project

Task ID	Description	Duration (days)		Cost	
		Minimum	Maximum	Minimum	Maximum
1	Excavation	1.5	2.5	600	1400
2	Shoring	0.5	1.0	300	500
3	Foundation	1.0	2.0	500	1100
4	Wall	1.0	2.0	500	1100
5	Backfilling	1.0	1.5	500	900

Table 9.2   Work breakdown details for the retaining wall project

Task ID	Crew ID	Segment ID	Quantity of work	Unit	Job condition Factor
1	1	1	0.8	$1000 \, m^3$	1.0
		2	0.8	$1000 \, m^3$	1.15
		3	0.8	$1000 \, m^3$	1.15
2	1	1	1.2	$100 \, m^2$	1.0
		2	1.0	$100 \, m^2$	1.0
		3	1.2	$100 \, m^2$	1.0
3	1	1	1.2	$100 \, m^3$	1.0
		2	1.2	$100 \, m^3$	1.0
		3	1.2	$100 \, m^3$	1.0
4	1	1	0.4	$100 \, m^3$	1.0
	2	2	0.4	$100 \, m^3$	1.0
	3	3	0.4	$100 \, m^3$	1.0
5	1	1	0.8	$1000 \, m^3$	1.0
	2	2	0.8	$1000 \, m^3$	1.0
	3	3	0.8	$1000 \, m^3$	1.0

breakdown details and the external logic of tasks are given Tables 9.2 and 9.3, respectively.

The minimum cost plan obtained by CONSCOM is shown in Figure 9.10. The duration of the project is 18.5 days. As a change

Table 9.3   External logic of tasks for the retaining wall project

Task	Predecessor	Relationship
Task 1		
Task 2, Crew 1, Segment 1	Task 1, Crew 1, Segment 1	FS, $L = 0.5$ days
Task 2, Crew 1, Segment 2	Task 1, Crew 1, Segment 2	FS, $L = 0.5$ days
Task 2, Crew 1, Segment 3	Task 1, Crew 1, Segment 3	FS, $L = 0.5$ days
Task 3, Crew 1, Segment 1	Task 2, Crew 1, Segment 1	FS, $L = 0$
Task 3, Crew 1, Segment 2	Task 2, Crew 1, Segment 2	FS, $L = 0$
Task 3, Crew 1, Segment 3	Task 2, Crew 1, Segment 3	FS, $L = 0$
Task 4, Crew 1, Segment 1	Task 3, Crew 1, Segment 1	FS, $L = 3$ days
Task 4, Crew 1, Segment 2	Task 3, Crew 1, Segment 2	FS, $L = 3$ days
Task 4, Crew 1, Segment 3	Task 3, Crew 1, Segment 3	FS, $L = 3$ days
Task 5, Crew 1, Segment 1	Task 4, Crew 1, Segment 1	FS, $L = 7$ days
Task 5, Crew 1, Segment 2	Task 4, Crew 1, Segment 2	FS, $L = 7$ days
Task 5, Crew 1, Segment 3	Task 4, Crew 1, Segment 3	FS, $L = 7$ days

order example, consider a situation where the soil conditions, after the contractor had started work on task 1, are found to be different from those assumed in design. The design is then revised to take into account the new information for the soil characteristics. Suppose the new quantities of work required for each segment of work of task 3 and 4 are now 150 and 50 m^3, respectively. Further, suppose the new design requires the construction of drains to remove excess soil water and reduce the pressure on the wall. Therefore, a new task (task 6) is added that starts after the work on the wall is completed. Task 6 has one crew that works consecutively over the three 10 m segments of the wall. The minimum and maximum durations required for completion of the work in one segment are 1 and 2 days, respectively, and the corresponding cost values are 500 and 300.

The owner has to decide on a new duration for the project. He/she must balance the cost of completing the work later than

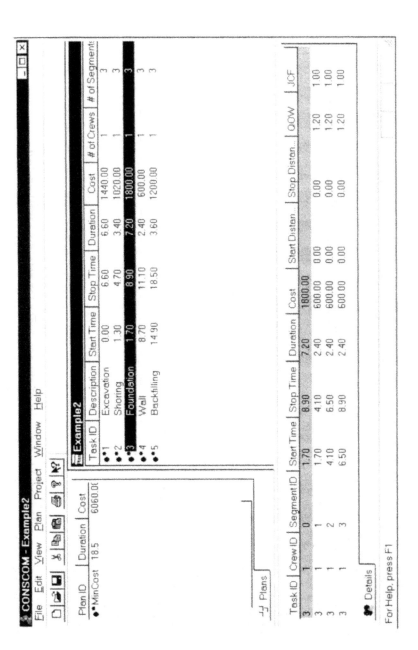

Figure 9.10 Minimum cost construction plan obtained by CONSCOM for retaining wall project

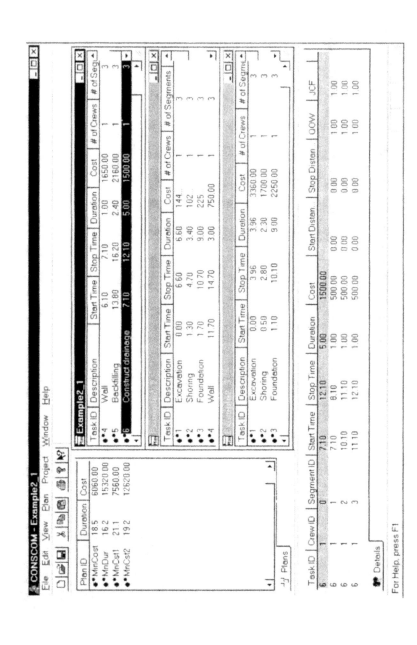

Figure 9.11 Analysis of change order for retaining wall project

initially anticipated with the cost of paying the contractor more to complete the job faster. This is therefore a cost-duration trade-off problem that the owner has to solve in order to make a final decision. CONSCOM allows the owner to quickly determine what the cost will be for a given duration of the project. Figure 9.11 shows CONSCOM's main window with the cost and duration values for the original minimum cost plan before the change order (ID: MinCost), and several updated plans after the change order including one with minimum completion duration (ID: MnDur), one with minimum cost without specifying any duration (ID: MnCst1), and a third plan with minimum cost for a given duration of 19 days (ID: MnCst2). It is assumed that the scheduling logic and the direct cost-duration functions are not varied in these examples. With such information the owner is then able to make an informed and intelligent decision regarding the cost and duration of the project and approve/disapprove the contractor's change order request.

9.7   CONCLUDING REMARKS

CONSCOM is an advanced new generation software system for scheduling and management of construction projects with cost optimization capability. It is particularly suitable for construction change order management. Change in a construction project after contract award is a common occurrence. It consumes a lot of resources in its implementation and in the resolution of any conflicts that often arise among the stakeholders. CONSCOM takes the owner's view of the problem. It aids the owner in schedule review, progress monitoring, and cost-time analyses for change order approval.

CONSCOM is based on an advanced integrated construction scheduling and cost model. Furthermore, it is developed using the latest software engineering techniques. And most importantly, it makes available all these features to the construction engineering professional in an easy to use, understand, and apply interface.

REGULARIZATION NEURAL NETWORK MODEL FOR CONSTRUCTION COST ESTIMATION[*]

10.1   INTRODUCTION

Estimating the cost of a construction project is an important task in construction project planning and management. Cost estimates are prepared and relied upon at all stages of the project's life cycle. They provide necessary information for decision making and effective control and management of construction projects. This task is currently performed by "experienced" construction cost estimators in a highly subjective manner. This approach, however, is subject to human errors and varying results depending on who the construction cost estimator is with possible litigation consequences.

Automating the process of construction cost estimation based on the subjective data is highly desirable not only for improving the

[*]This chapter is based on the article: Adeli, H. and Wu, M. (1998), "Regularization Neural Network for Construction Cost Estimation," *Journal of Construction Engineering and Management*, ASCE, 124(1), pp. 18–24, and is reproduced by permission of the publisher (ASCE).

efficiency but also for removing the subjective questionable human factors as much as possible. The problem is not amenable to traditional problem solving approaches. The costs of construction material, equipment, and labor depend on numerous factors with no explicit mathematical model or rule for price prediction. Recently, neural networks have been used for learning and prediction problems with no explicit models such as securities price prediction (Hutchinson *et al.*, 1994).

Learning from previous data or examples, neural networks can make very reasonable estimations without using specific experts and rules. As an example, the price of a concrete pavement is influenced by a number of factors including the quantity, the dimension (thickness of the pavement), the local economic factors, and the time of the construction. The problem to be investigated is whether using the values for these factors obtained from the past construction projects a neural network model can estimate the price of a future construction project accurately.

In this chapter, first the concepts of estimation, learning, and noisy curve fitting are described and formulated mathematically. Next, a special case of radial-based function neural networks, called regularization neural network, is formulated for estimating the cost of construction projects. Then, the model is applied to reinforced concrete pavement cost estimation as an example.

10.2 ESTIMATION, LEARNING AND NOISY CURVE FITTING

The most fundamental problem that neural networks have been used to solve is learning. But, it is very difficult, if not impossible, to present a precise definition of learning. In order to model learning computationally, however, it has to be defined in a pragmatic manner rather than as an abstract concept. Learning can be defined as a self-organizing process, a mapping process, an optimization process, or a decision making process. The last definition is based on the observation that given a set of examples and a stimulus one makes a decision about the response to the stimulus.

Consider the special case of supervised learning where the system is first trained by a set of input–output examples. Then, given a new input the learner decides the output. This is an ill-posed problem because the answer can be any multitude of values. The selected answer depends on the generalization criteria or constraints chosen for the decision process. The advantage of viewing the learning process as a decision making process is its explicit representation of the generalization criteria.

An estimation problem can be formulated as a supervised learning problem. Consider a one-input–one-output noisy system with input x and output y. The system can be expressed mathematically as

$$y = f(x) + e \tag{10.1}$$

where $f(x)$ is a function of the input variable x and e is an independent noise function with zero mean. A set of input–output examples x_i and y_i ($i = 1, 2,..., N$) is given. The estimation problem is: for any given input x find the best estimate of the output y. The best estimate can be defined as the estimate that minimizes the average error between the actual and estimated outputs. Thus, such a supervised learning problem becomes equivalent to a mapping or curve-fitting problem in a multi-dimensional hyperspace. It must be pointed out that the traditional statistical approaches to curve fitting such as the regression analysis fail to represent problems with no explicit mathematical model accurately in a multi-dimensional space of variables. The neural network approach, on the other hand, can solve such problems more effectively.

Figure 10.1 shows a very simple example of curve fitting and learning. The dots represent the example data points that include noise. The dashed line represents the properly learned curve. The solid line represents the over-fitted learned curve. For this curve, the training error is very small (because the learned curve passes through all the training data points), but the estimation or generalization error is large. This is due to the fact that the influence of the noise has not been taken into account at all. This problem is referenced to as overfitting which leads to less than satisfactory learning. Avoiding the overfitting problem is very important for accurate estimation and learning.

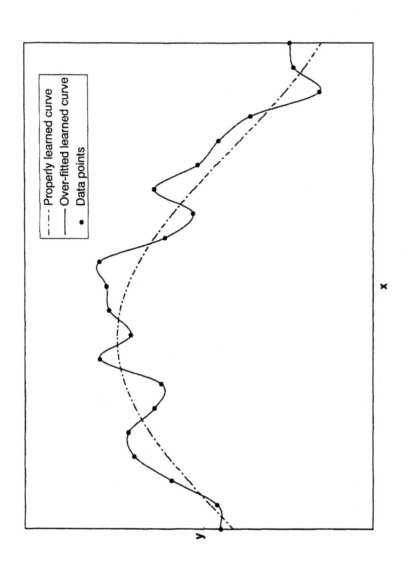

Figure 10.1 Comparison of properly learned and over-fitted learned curve

A mathematical definition of learning is now formulated as a mapping (generalization of curve fitting) problem in a multi-dimensional hyperspace. A neural network is designed to perform a nonlinear mapping function s from a p-dimensional input space R^p to a one-dimensional output space R^1.

$$s : R^p \to R^1 \tag{10.2}$$

The set of N available input–output data can be described as

Input signal: $\qquad\qquad \mathbf{x}_i \in R^p \quad i = 1, N$

Example out put signal: $\quad d_i \in R^1 \quad i = 1, N$

where $\mathbf{x}_i = \{x_1^i, x_2^i, ..., x_p^i\}$ is the ith example with p input attributes (x_n^i is the nth attribute of the ith example) and d_i is the corresponding example output. The approximation mapping function is denoted by $F(\mathbf{x})$.

What is the best fit? This is an important question. Because of the existence of the noise in the data examples, a perfect fit, that is when $F(\mathbf{x}_i) = d_i$, usually is not the best fit. In this case, the approximation function is often very curvy with numerous steep peaks and valleys which leads to poor generalization. This is the overfitting problem mentioned earlier. Two other fitting situations can also be recognized: underfitting with over-smooth surfaces resulting in poor generalization, and proper fitting. Only the last type of fitting can lead to accurate generalization and estimation and this is the research challenge. A method to achieve proper fitting will be discussed in the following sections.

Highway construction costs are affected by many factors, but only a few main factors are usually recorded and can be considered in the mathematical modeling of the cost estimation problem. As such, the highway construction data are very noisy and the noise is the result of many unpredictable factors such as human judgement factors, random market fluctuations, and weather conditions. Consequently, finding a properly fitted approximation is extremely important. Otherwise, the predicted cost will have a substantial error.

One approach to solve this problem is the multilayer feedforward backpropagation neural network (Haykin, 1999). The problem with this approach is that the generalization properties (underfitted, overfitted, or properly fitted) depend on many factors including the architecture (number of hidden layers and number of nodes in the hidden layers), initial weights of the links connecting the nodes, and number of iterations for training the system. The performance of the algorithm depends highly on the selection of these parameters. The problem of arbitrary trial-and-error selection of the learning and momentum ratios encountered in the momentum backpropagation learning algorithm can be circumvented by the adaptive conjugate gradient learning algorithm developed by Adeli and Hung (1994). But, that algorithm does not address the issue of noise in the data. In the highway construction estimation problem, the data has substantial noise and the neural network algorithm must be able to address the issue of noise in the data properly. For this purpose we employ a neural network architecture called regularization network to obtain properly fitted approximation function and solve the construction estimation problem accurately.

10.3   REGULARIZATION NETWORKS

According to the regularization theory (Tikhonov and Arsenin, 1977; Haykin, 1999), the approximation mapping function F is determined by minimizing an error function $E(F)$, consisting of two terms in the following form:

$$E(F) = E_s(F) + E_c(F) \tag{10.3}$$

The first term is the standard error term measuring the error between the sample example response d_i and the corresponding computed response o_i and is defined as follows:

$$E_s(F) = \frac{1}{2}\sum_{i=1}^{N}(d_i - o_i)^2 = \frac{1}{2}\sum_{i=1}^{N}[d_i - F(\mathbf{x}_i)]^2 \tag{10.4}$$

As discussed above, the perfect fit may not be the best answer due to the noise in the data. In order to overcome the overfitting problem a regularization term is added to the standard error term whose function is to smoothen the approximation function. This term is defined as

$$E_c(F) = \frac{1}{2}\|PF\|^2 \tag{10.5}$$

where the symbol $\|g\|$ denotes the norm of function $g(\mathbf{x})$ defined as:

$$\|g\| = \int_{R^p} [g(\mathbf{x})]^2 dx_1 dx_2 \cdots dx_p \tag{10.6}$$

and P is a linear differential operator defined as (Poggio and Girosi, 1990; Al-Gwaiz, 1992):

$$\|PF\|^2 = \sum_{k=0}^{K} b_k \|D^k F(\mathbf{x})\|^2 \tag{10.7}$$

In Eq. (10.7), K is a positive integer, b_k's ($k = 0, 1,...K$) are positive real numbers, and the norm of the differential operator D^k, is defined as:

$$\|D^k F\|^2 = \sum_{|\alpha|=k} \int_{R^p} [\partial^\alpha F(\mathbf{x})]^2 dx_1 dx_2 \cdots dx_p \tag{10.8}$$

The multi-index $\alpha = \{\alpha_1, \alpha_2, ..., \alpha_p\}$ is a sequence of non-negative integers whose order is defined as $|\alpha| = \sum_{i=1}^{p} \alpha_i$. In Eq. (10.8), the partial differential term inside the bracket is defined as

$$\partial^\alpha F(\mathbf{x}) = \frac{\partial^{|\alpha|} F(\mathbf{x})}{\partial x_1^{\alpha_1} \partial x_2^{\alpha_2} \cdots \partial x_p^{\alpha_p}} \tag{10.9}$$

Therefore, the regularization term is

$$E_c(F) = \frac{1}{2} \sum_{k=0}^{K} \sum_{|\alpha|=k} \int_{R^p} b_k [\partial^\alpha F(\mathbf{x})]^2 dx_1 dx_2 \cdots dx_p \tag{10.10}$$

This function is simply the summation of the integrations of the partial derivatives of the approximation function squared. As such, the regularization term is small when the function is smooth because the derivatives tend to be small and vice versa.

For $b_k = \dfrac{\beta^{2k}}{K!2^k}$ and K approaching infinity, where β is a positive real number, it can be proven that by minimizing the error function, Eq. (10.3), with respect to the approximation function, the solution of the problem can be written in the following form (Poggio and Girosi, 1990):

$$F(\mathbf{x}) = \sum_{i=1}^{N} w_i \exp\left(-\frac{1}{2\beta^2}\|\mathbf{x} - \mathbf{x}_i\|^2\right) = \sum_{i=1}^{N} w_i \exp\left(-\sigma\|\mathbf{x} - \mathbf{x}_i\|^2\right) \quad (10.11)$$

where $\sigma = \dfrac{1}{2\beta^2}$ and w_i's are real numbers. Eq. (10.11) is a linear superposition of multivariate Gaussian functions with centers $\mathbf{x}_i$'s located at the data points.

The architecture of the regularization network for construction cost estimation is shown schematically in Figure 10.2. It consists of an input layer, a hidden layer, and an output layer. The number of nodes in the input layer is equal to p, the number of input attributes. The number of the nodes in the hidden layer is equal to N, the number of the training examples. The output node gives the estimated construction cost.

The network shown in Figure 10.2 is a feedforward network. The nodes in the hidden layer represent the nonlinear multivariate Gaussian activation functions $G_i(\mathbf{x}) = \exp\left(-\varpi\|\mathbf{x} - \mathbf{x}_i\|^2\right)$. In other

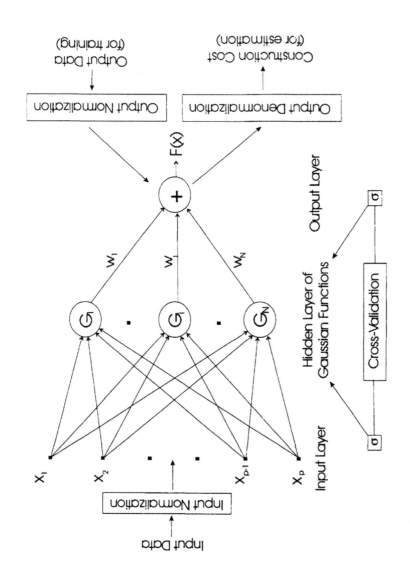

Figure 10.2 Architecture of regularization network for construction cost estimation

works, the output of the ith node in the hidden layer is $G_i(\mathbf{x}) = \exp\left(-\varpi \|\mathbf{x} - \mathbf{x}_i\|^2\right)$. The input and hidden layers are fully connected. That means every node in the hidden layer receives inputs from all the nodes in the input layer. The links connecting the hidden layer to the output layer represent the weights w_i's in the approximation function, Eq. (10.11).

The learning process of the regularization network consists of two steps. In the first step, the value of the parameter σ in Eq. (10.11) is found by a cross-validation procedure to be described in Section 10.5. The smoothness of the approximation function is primarily controlled by this parameter. The smaller the value of σ, the smoother the approximation function will be. We will call σ the smoothing parameter. In the second step, w_i's are found using the method described in next section.

10.4   DETERMINATION OF WEIGHTS OF REGULARIZATION NETWORK

The smoothing parameter σ and the weights w_i's depend on each other in a complicated way and consequently must be calculated iteratively. In this section, a method is presented for finding the weights w_i's. Defining the following matrices

$$\mathbf{d} = [d_1, d_2, ..., d_N]^T \tag{10.12}$$

$$\mathbf{G} = \begin{bmatrix} G(\mathbf{x}_1;\mathbf{x}_1) & G(\mathbf{x}_1;\mathbf{x}_2) & \cdots & G(\mathbf{x}_1;\mathbf{x}_N) \\ G(\mathbf{x}_2;\mathbf{x}_1) & G(\mathbf{x}_2;\mathbf{x}_2) & \cdots & G(\mathbf{x}_2;\mathbf{x}_N) \\ \vdots & \vdots & & \vdots \\ G(\mathbf{x}_N;\mathbf{x}_1) & G(\mathbf{x}_N;\mathbf{x}_2) & \cdots & G(\mathbf{x}_N;\mathbf{x}_N) \end{bmatrix} \tag{10.13}$$

$$\mathbf{w} = \begin{bmatrix} w_1, w_2, \ldots, w_N \end{bmatrix}^T \tag{10.14}$$

where $G(\mathbf{x}_i;\mathbf{x}_j) = \exp\left(-\sigma\left\|\mathbf{x}_i - \mathbf{x}_j\right\|^2\right)$. It can be shown that the solution of the regularization problem, i.e., the weights w_i's, satisfies the following equation (Haykin, 1999):

$$(\mathbf{G}+\mathbf{I})\mathbf{w} = \mathbf{d} \tag{10.15}$$

where $\mathbf{I}$ is the $N{\times}N$ identity matrix. If the matrix $(\mathbf{G}+\mathbf{I})$ is not ill-conditioned, the solution of the linear equation represented by Eq. (10.15) can be solved by a linear equation solver such as the Gauss-Jordan elimination or LU decomposition method. The $N{\times}N$ matrix $(\mathbf{G}+\mathbf{I})$, however, is large and usually suffers from numerical ill-conditioning. This leads to zero pivot in the Gauss-Jordan elimination method resulting in large errors in the solution. The aforementioned approach was in fact employed in this research but without any success. The elements of the optimum vector $\mathbf{w}$ were found to be very large due to numerical instability.

To overcome the numerical ill-conditioning problem, a singular value decomposition method is used to find the weights w_i's (Press et al., 1988). In this approach, the matrix $(\mathbf{G}+\mathbf{I})$ is first decomposed as

$$\mathbf{G}+\mathbf{I} = \mathbf{U}\mathbf{C}\mathbf{V}^T \tag{10.16}$$

where $\mathbf{U}$ and $\mathbf{V}$ are NxN orthonormal matrices (i.e. $\mathbf{UU}^T = \mathbf{I}$ and $\mathbf{VV}^T = \mathbf{I}$) and $\mathbf{C}$ is an NxN diagonal matrix with diagonal entries c_i's, called singular values, where $|c_1| \geq |c_2| \geq \cdots \geq |c_N|$.

Having found the matrices, $\mathbf{U}$, $\mathbf{V}$, and $\mathbf{C}$, the weight vector can be found from

$$\mathbf{w} = \sum_{i=1}^{J} \left(\frac{\mathbf{U}_{(i)} \cdot \mathbf{d}}{c_i} \right) \mathbf{V}_{(i)} \qquad (10.17)$$

where $\mathbf{U}_{(i)}$, $i = 1,N$ denotes the ith column of $\mathbf{U}$ and $\mathbf{V}_{(i)}$, $i = 1,N$ denotes the ith column of $\mathbf{V}$. In Eq. (10.17), the summation is performed over J terms and not the entire N terms in order to avoid numerical ill-conditioning due to division by very small numbers and the truncation error. The terms c_i's in the denominator of Eq. (10.17) cannot take very small values. All of the singular values used in Eq. (10.17) are greater than $\varepsilon = N \varepsilon_m |c_1|$ where ε_m is the machine precision, and c_1 is the largest singular value. By selecting a predetermined value for ε_m the small values of c_i's are excluded from the summation in Eq. (10.17), and the summation is done over J terms such that $|c_i| > \varepsilon$ for any $i \leq J$ and $|c_i| \leq \varepsilon$ for any $i > J$.

10.5   PROPER GENERALIZATION AND ESTIMATION BY CROSS-VALIDATION

For proper solution of the problem and accurate estimation, a trade-off is necessary between minimizing the standard error term, Eq.

(10.4), and smoothness of the approximation function. As mentioned earlier, the smoothness of the approximation function and the generalization properties of the network are influenced by the parameter σ. In order to obtain a proper value for σ a method used in statistical pattern recognition called cross-validation is employed in this research (Fukunaga, 1990).

In the cross-validation method, the available set of examples is divided randomly into two sets, a training set $\{x_t^n, d_t^n\}$ $n = 1, 2, ..., N_t$, and a validation set $\{x_v^n, d_v^n\}$ $n = 1, 2, ..., N_v$, where subscripts t and v refer to training and validation sets, respectively, and N_t and N_v are the numbers of training and validation examples, respectively. The network is trained with the training set $\{x_t^n, d_t^n\}$ $n = 1, 2, ..., N_t$ using different values of σ within a given range. This range is problem-dependent and is determined by experience and numerical experimentation.

For each value of σ, the weights w_i's are found using Eqs. (10.16) and (10.17) and an average training error is calculated in the following form:

$$E_t = \sqrt{\sum_{n=1}^{N_t} [d_t^n - F(x_t^n)]^2 \Big/ N_t} \qquad (10.18)$$

Next, using the validation set $\{x_v^n, d_v^n\}$ $n = 1, 2, ..., N_v$ an average validation error is calculated in the following form:

$$E_v = \sqrt{\sum_{n=1}^{N_v} \left[d_v^n - F(\mathbf{x}_v^n) \right]^2 \Big/ N_v} \qquad (10.19)$$

Typical trend relationships between the average training and validation errors and the smoothing factor σ are shown conceptually in Figure 10.3. The average training error always decreases with an increase in the magnitude of σ for a numerically stable algorithm. In contrast, the average validation error curve does not have a continuously decreasing trend. Rather, one can identify a minimum on this curve. As mentioned earlier in the chapter, broadly speaking a large σ indicates overfitting and a small σ indicates underfitting. The σ corresponding to the global minimum point on the validation curve represents the properly fitted estimation curve. The average validation error gives an estimate of the estimation/prediction error.

10.6   INPUT AND OUTPUT NORMALIZATION

Because the regularization network uses spherically symmetric multivariate Gaussian activation functions, to improve the performance the input variables are normalized so that they span similar ranges in the input space. The purpose of the normalization is to ensure that all examples in the training set have similar influence in the learned approximation function. In other words, it is statistically desirable to have variables with zero mean and the same unit standard deviation. This can be achieved by using the following change of variables and normalization procedure (Fukunaga, 1990):

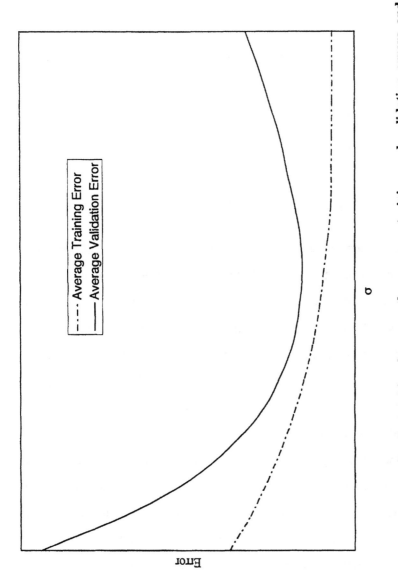

Figure 10.3 Typical trend relationships between the average training and validation errors and the smoothing factor σ

$$\tilde{x}_i^n = \frac{x_i^n - \bar{x}_i}{\sigma_i} \tag{10.20}$$

where $\tilde{x}_i^n$, $i = 1,2,...,N$ are the normalized input data, and

$$\bar{x}_i = \frac{1}{N} \sum_{n=1}^{N} x_i^n \quad i = 1,2,...,p \tag{10.21}$$

and

$$\sigma_i^2 = \frac{1}{N-1} \sum_{n=1}^{N} \left(x_i^n - \bar{x}_i\right)^2 \quad i = 1,2,...,N \tag{10.22}$$

are the means and standard deviations of the original set of variables.

The Gaussian activation function is maximum at its center (data point) and approaches zero at large distances from the center. In other words, statistically speaking, the use of the Gaussian activation function amounts to large output near the center (data point) and zero output at large distances from the center where there is no data point. But, the lack of data point does not necessarily mean the output is zero at large distances from the available sample data points.

One may argue that it is not possible to make an accurate estimate at large distances from the example data points. This is the well-known extrapolation problem. While the regularization theory solves the interpolation problem accurately it is not concerned with the extrapolation problem. But, a practical estimation system should not fail at the boundaries of the available data domain abruptly. Consequently, to improve the estimation accuracy at large distances from the available data points, first a linear trend (hyperplane) is found through the example data points by performing a linear regression analysis. Next, the output data are normalized with respect to this hyperplane (the outputs are measured from this plane instead of a zero-base hyperplane). Finally, regularization network is applied using the normalized data output. This process will bring the estimates at large distances from the available data points close to the linear trend hyperplane.

Mathematically, the function $\sum_{n=1}^{N}\left(d_i^n - \sum_{i=1}^{p} a_i \tilde{x}_i^n - a_0\right)^2$ is minimized with respect to linear parameters a_i's $(i = 0, 1, 2, \ldots, p)$ in order to find the linear trend hyperplane. This hyperplane is represented by

$$y = \sum_{i=1}^{p} a_i \tilde{x}_i + a_0 \tag{10.23}$$

Then, the normalized output data redefined as

$$\tilde{d}_i^n = d_i^n - \sum_{i=1}^{p} a_i \tilde{x}_i^n - a_0 \tag{10.24}$$

10.7   APPLICATION

The computational model presented in this chapter has been implemented in the programming language MATLAB (MathWorks, 1992) and applied to the problem of estimating the cost of concrete pavements. The reason for selection of MATLAB is the availability of a large number of built-in numerical analysis functions such as singular value decomposition.

The data set was collected from the files of previous projects at the Ohio Department of Transportation. It includes 242 examples of construction costs for reinforced concrete pavements. The cost factors used in the examples are the quantity and the dimension (thickness) of the pavement.

10.7.1 Example 1

In this example only the quantity information is used. The variation of the unit cost versus the quantity is presented in Figure 10.4. Because the reinforced concrete pavement quantity has a large variation the quantity scale in Figure 10.4 is a logarithmic scale. Figure 10.4 shows clearly that the highway construction cost data are very noisy.

The training set of 121 examples and the validation set of 121 examples are selected randomly from the available 242 data examples. Variations of the average training and validation errors with respect to the smoothing parameter σ are shown in Figure 10.5. The trend for both curves is similar to trends discussed in Section 10.5 and Figure 10.3. The minimum point on the average validation error curve corresponds to $\sigma = 0.8$ which represents the value needed for the proper generalization. The corresponding average training and validation errors for the unit cost of the concrete pavement are $6.45/m^3$ and $7.22/m^3$, respectively. For comparison, the average unit cost of the concrete pavement for the 242 example data is

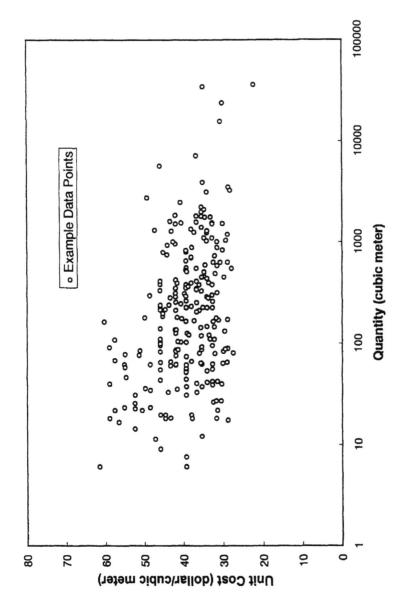

Figure 10.4 Unit cost versus quantity for the example data set

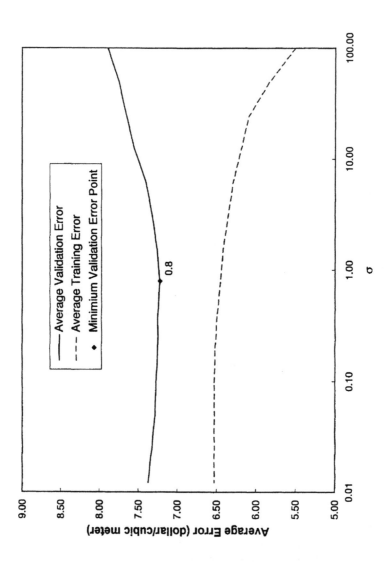

Figure 10.5 The average training and validation errors for different values of α using only quantity information

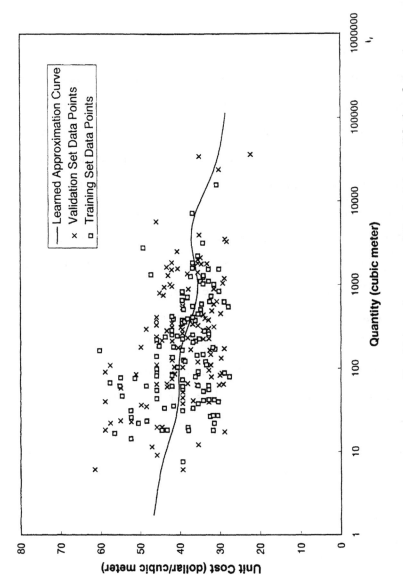

Figure 10.6 Proper generalized learned curve and the training/validation data set

$39.2/\text{m}^3$. Figure 10.6 shows the learned curve along with training and validation data sets.

10.7.2 Example 2

In this example, the quantity and the dimension (thickness of the pavement) are used as input attributes. The unit cost versus the concrete pavement thickness for the 242 example data are shown in Figure 10.7. Similar to example 1, the data set is divided into 121 training and 121 validation examples. Variations of the average training and validation errors with respect to the smoothing parameter σ are shown in Figure 10.8. The smoothing parameter corresponding to the minimum point on the average validation error curve is found to be $\sigma = 0.05$. The corresponding average training and validation errors for the unit cost of the concrete pavement are $6.3/\text{m}^3$ and $6.7/\text{m}^3$, respectively. The learned approximation function is a surface in a three-dimensional space. The average validation error of this example is less than that of Example 1. As the number of attributes is increased the average validation error decreases which means the construction cost is estimated more accurately.

10.8 CONCLUSION

In this chapter, a regularization neural network was presented for estimating the cost of construction projects. The problem is formulated in terms of an error function consisting of a standard error term and a regularization term. The purpose of the latter term is

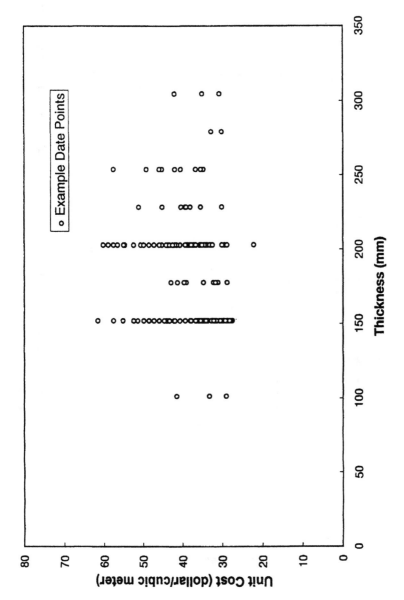

Figure 10.7 Unit cost versus dimension for the example data set

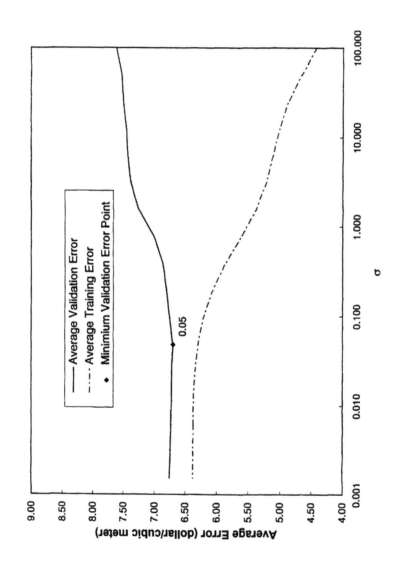

Figure 10.8 The average training and validation errors for different values of σ using quantity and dimension information

to compensate for the overfitting problem and to improve the cost estimation outside of the available data points.

The traditional regression analysis methods can fit the data only in certain types of functions such as polynomial functions. Further, a major assumption is made that data must fit one of these functions. In regularization neural network approach presented in this chapter, on the other hand, no assumption is made about the shape of the approximation function to be learned. The only assumptions made are the continuity and the general smoothness of the function.

The neural network model presented in this chapter has the following major advantages over other neural networks algorithms such as the backpropagation neural network:

- The regularization neural network is based on a solid mathematical foundation. This makes the cost estimation model consistently reliable and predictable.

- The result of estimation from the regularization neural network depends only on the training examples. It does not depend on the architecture of the neural network (such as the number of nodes in the hidden layer), the learning parameters (such as the learning and momentum ratios in the BP algorithm), and the number of iterations required for training the system. As such, it can be said the regularization neural network presented in this chapter is an objective cost estimator.

- The problem of noise in the data, which is important in the highway construction cost data, is taken into account in a rational manner.

The generalization error of the regularization networks can be attributed to insufficient data examples which can be improved by increasing the database of examples from previous construction projects and intrinsic noise due to non-quantifiable and unpredictable factors which is impossible to avoid.

BIBLIOGRAPHY

Abd. Majid, M.Z. and McCaffer, R. (1998), "Factors of Non-Excusable Delays that Influence Contractors' Performance," *Journal of Management in Engineering*, ASCE, 14(3), pp. 42–49.

Abdalla, K.M. and Stavroulakis, G.E. (1995), "A Backpropagation Neural Network Model for Semi-rigid Steel Connections," *Microcomputers in Civil Engineering*, 10(2), pp. 77–87.

Adeli, H., Ed. (1988), *Expert Systems in Construction and Structural Engineering*, Chapman and Hall, London.

Adeli, H., Ed. (1990a), *Knowledge Engineering – Vol. One – Fundamentals*, McGraw-Hill, New York.

Adeli, H., Ed. (1990b), *Knowledge Engineering – Vol. Two – Applications*, McGraw-Hill, New York.

Adeli, H., Ed. (1992a), *Parallel Processing in Computational Mechanics*, Marcel Dekker, New York.

Adeli, H., Ed. (1992b), *Supercomputing in Engineering Analysis*, Marcel Dekker, New York.

Adeli, H., Ed. (1994), *Advances in Design Optimization*, E. & F.N. Spon, London.

Adeli, H. (1999), "Competitive Edge and Environmentally-conscious Design Through Concurrent Engineering," *Assembly Automation*, 19(2), pp. 92–94.

Adeli, H. (2000), "Concurrent Engineering in the Construction Industry," in Ghodous, P. and Vandorpe, D., Eds., *Advances in Concurrent Engineering*, Technomic Publishing Co., Lancaster, PA.

Adeli, H. (2001), "Neural Networks in Civil Engineering – 1989–2000," *Computer-Aided Civil and Infrastructure Engineering*, 16(2).

Adeli, H. and Balasubramanyam, K.V. (1988), *Expert Systems for Structural Design – A New Generation*, Prentice-Hall, Englewood Cliffs, New Jersey.

Adeli, H. and Hung, S.L. (1993), "A Fuzzy Neural Network Learning Model for Image Recognition," *Integrated Computer-Aided Engineering*, 1(1), pp. 43–55.

Adeli, H. and Hung, S.L. (1994), "An Adaptive Conjugate Gradient Learning Algorithm for Efficient Training of Neural Networks," *Applied Mathematics and Computations*, 62(1), pp. 81–102.

Adeli, H. and Hung, S.L. (1995), *Machine Learning – Neural Networks, Genetic Algorithms, and Fuzzy Systems*, John Wiley and Sons, New York.

Adeli, H. and Kamal, O. (1993), *Parallel Processing in Structural Engineering*, Elsevier, London.

Adeli, H. and Kao, W.-M. (1996), "Object-Oriented Blackboard Models for Integrated Design of Steel Structures," *Computers and Structures*, 61(3), pp. 545–561.

Adeli, H. and Karim, A. (1997a), "Neural Dynamics Model for Optimization of Cold-Formed Steel Beams," *Journal of Structural Engineering*, ASCE, 123(11), pp. 1535–1543.

Adeli, H. and Karim, A. (1997b), "Scheduling/Cost Optimization and Neural Dynamics Model for Construction," *Journal of Construction Engineering and Management*, ASCE, 123(4), pp. 450–458.

Adeli, H. and Karim, A. (2000), "Fuzzy-Wavelet RBFNN Model for Freeway Incident Detection," *Journal of Transportation Engineering*, ASCE, 126(6), pp. 464–471.

Adeli, H. and Kumar, S. (1999), *Distributed Computer-Aided Engineering for Analysis, Design, and Visualization*, CRC Press, Boca Raton, Florida.

Adeli, H. and Park, H.S. (1995a), "A Neural Dynamics Model for Structural Optimization – Theory," *Computers and Structures*, 57(1), pp. 383–390.

Adeli, H. and Park, H.S. (1995b), "Optimization of Space Structures by Neural Dynamics Model," *Neural Networks*, 8(5), pp. 769–781.

Adeli, H. and Park, H.S. (1995c), "Counterpropagation Neural Networks in Structural Engineering," *Journal of Structural Engineering*, ASCE, 121(8), pp. 1205–1212.

Adeli, H. and Park, H.S. (1996a), "Hybrid CPN-Neural Dynamics Model for Discrete Optimization of Steel Structures," *Microcomputers in Civil Engineering*, 11(5), pp. 355–366.

Adeli, H. and Park, H.S. (1998), *Neurocomputing for Design Automation*, CRC Press, Boca Raton, Florida.

Adeli, H. and Saleh, A. (1999), *Control, Optimization, and Smart Structures – High-Performance Bridges and Buildings of the Future*, John Wiley & Sons, New York.

Adeli, H. and Samant, A. (2000), "An Adaptive Conjugate Gradient Neural Network-Wavelet Model for Traffic Incident Detection," *Computer-Aided Civil and Infrastructure Engineering*, 15(4), pp. 251–260.

Adeli, H. and Soegiarso, R. (1999), *High-Performance Computing in Structural Engineering*, CRC Press, Boca Raton, Florida.

Adeli, H. and Wu, M. (1998), "Regularization Neural Network for Construction Cost Estimation," *Journal of Construction Engineering and Management*, ASCE, 124(1), pp. 18–24.

Adeli, H. and Yeh, C. (1989), "Perceptron Learning in Engineering Design," *Microcomputers in Civil Engineering*, 4(4), pp. 247–256.

Adeli, H. and Yu, G. (1993a), "An Object-Oriented Data Management Model for Numerical Analysis in Computer-Aided Engineering," *Microcomputers in Civil Engineering*, 8(3), pp. 199–209.

Adeli, H. and Yu, G. (1993b), "Concurrent OOP Model for Computer-Aided Engineering using Blackboard Architecture," *Journal of Parallel Algorithms and Applications*, 1(2), pp. 315–337.

Adeli, H. and Zhang, J. (1993), "An Improved Perceptron Learning Algorithm," *Neural, Parallel, and Scientific Computations*, 1(2), pp. 141–152.

AISC (1995), *Manual of Steel Construction – Allowable Stress Design*, American Institute of Steel Construction, 9th edition, 2nd revision, Chicago, Illinois.

AISC (1998), *Manual of Steel Construction – Load and Resistance Factor Design*, Vol. I, 2nd ed., 2nd revision, American Institute of Steel Construction, Chicago, Illinois.

AISI (1968), *Specification for the Design of Cold-Formed Steel Structural Members*, American Iron and Steel Institute, Washington, DC.

AISI (1989), *Specification for the Design of Cold-Formed Steel Structural Members*, 1986 ed. with 1989 Addendum, American Iron and Steel Institute, Washington, DC.

AISI (1991), *Load and Resistance Factor Design Specification for Cold-Formed Steel Structural Members*, American Iron and Steel Institute, Washington, DC.

AISI (1996), *Specification for The Design of Cold-Formed Steel Structural Members*, American Iron and Steel Institute, Washington, D.C.

AISI (1997), *Cold-Formed Steel Design Manual*, American Iron and Steel Institute, Washington, D.C.

Aleksander, I. and Morton, H. (1993), *Neurons and Symbols – The Stuff That Mind is Made Of*, Chapman and Hall, London.

Al-Gwaiz, M.A. (1992), *Theory of Distributions*, Marcel Dekker, New York.

Alsugair, A.M. and Al-Qudrah, A.A. (1998), "Artificial Neural Network Approach for Pavement Maintenance," *Journal of Computing in Civil Engineering*, ASCE, 12(4), pp. 249–255.

Amin, S.M., Rodin, E.Y., and Garcia-Ortiz, A. (1998), "Traffic Prediction and Management via RBF Neural Nets and Semantic Control," *Computer-Aided Civil and Infrastructure Engineering*, 13(5), pp. 315–327.

Anantha Ramu, S. and Johnson, V.T. (1995), "Damage Assessment of Composite Structures – A Fuzzy Logic Integrated Neural Network Approach," *Computers and Structures*, 57(3), pp. 491–502.

Anderson, D., Hines, E.L., Arthur, S.J. and Eiap, E.L. (1997), "Application of Artificial Neural Networks to the Prediction of Minor Axis Steel Connections," *Computers and Structures*, 63(4), pp. 685–692.

Anderson, J.A. (1995), *An Introduction to Neural Networks*, A Bradford Book, MIT Press, Cambridge, Massachusetts.

Ankireddi, S. and Yang, H.T.Y. (1999), "Neural Networks for Sensor Fault Correction in Structural Control," *Journal of Structural Engineering*, 125(9), pp. 1056–1064.

Arbib, M.A. (1995), *The Handbook of Brain Theory and Neural Networks*, MIT Press, Cambridge, Massachusetts.

Arditi, D., Oksay, F.E. and Tokdemir, O.B. (1998), "Predicting the Outcome of Construction Litigation Using Neural Networks," *Computer-Aided Civil and Infrastructure Engineering*, 13(2), pp. 75–81.

Arditi, D. and Tokdemir, O.B. (1999), "Comparison of Case-Based Reasoning and Artificial Neural Networks," *Journal of Computing in Civil Engineering*, ASCE, 13(3), pp. 162–169.

Arslan, M.A. and Hajela, P. (1997), "Counterpropagation Neural Networks in Decomposition Based Optimal Design," *Computers and Structures*, 65(5), pp. 641–650.

ASTM (1996), *1996 Annual Book of ASTM Standards*, American Society for Testing and Materials, West Conshohocken, PA.

Attoh-Okine, N.O. (2001), "Grouping Pavement Condition Variables for Performance Modeling Using Self Organizing Maps," *Computer-Aided Civil and Infrastructure Engineering*, 16(2)

Bahreininejad, A., Topping, B. H. V. and Khan, A. I. (1996), "Finite Element Mesh Partitioning Using Neural Networks," *Advances in Engineering Software*, 27, pp. 103–115.

Bani-Hani, K. and Ghaboussi, J. (1998), "Nonlinear Structural Control Using Neural Networks," *Journal of Engineering Mechanics*, 124(3), pp. 319–327.

Barai, S.V. and Pandey, P.C. (1995), "Vibration Signature Analysis Using Artificial Neural Networks," *Journal of Computing in Civil Engineering*, ASCE, 9(4), pp. 259–265.

Basheer, I.A. (2000), "Selection of Methodology for Neural Network Modeling of Constitutive Hysteresis Behavior of Soils," *Computer-Aided Civil and Infrastructure Engineering*, 15(6), pp. 445–463.

Basheer, I.A. and Najjar, Y.M. (1996), "Predicting Dynamic Response of Adsorption Columns with Neural Nets," *Journal of Computing in Civil Engineering*, ASCE, 10(1), pp. 31–39.

Baumer, D., Gryczan, G., Knoll, R., Lilienthal, C., Riehle, D. and Zullighoven, H. (1997), "Framework Development for Large Systems," *Communications of the ACM*, 40(10), pp. 52–59.

Berke, L., Patnaik, S.N. and Murthy, P.L.N. (1993), "Optimum Design of Aerospace Structural Components Using Neural Networks," *Computers and Structures*, 48(6), pp. 1001–1010.

Biedermann, J.D. (1997), "Representing Design Knowledge with Neural Networks," *Computer-Aided Civil and Infrastructure Engineering*, 12(4), pp. 277–285.

Bishop, C.M. (1995), *Neural Networks for Pattern Recognition*, *Clarendon Press*, Oxford, UK.

Brown, D.A., Murthy, P.L.N. and Berke, L. (1991), "Computational Simulation of Composite Ply Michromechanics Using Artificial Neural Networks," *Microcomputers in Civil Engineering*, 6(2), pp. 87–97.

Cal, Y. (1995), "Soil Classification by Neural Network," *Advances in Engineering Software*, 22, pp. 95–97.

Cao, X., Sugiyama, Y. and Mitsui, Y. (1998), "Application of Artificial Neural Networks to Load Identification," *Computers and Structures*, 69, pp. 63–78.

Carpenter, G.A., Grossberg, S. and Rosen, D.B. (1991), "Fuzzy ART: Fast Stable Learning of Analog Patterns by an Adaptive Resonance System," *Neural Networks*, 4, pp. 759–771.

Carpenter, W.C. and Barthelemy, J.-F. (1994), "Common Misconceptions About Neural Networks as Approximators," *Journal of Computing in Civil Engineering*, ASCE, 8(3), pp. 345–358.

Castillo, E., Cobo, A., Gutierrez, J.M. and Pruneda, E. (2000a), "Functional Networks – A New Network-Based Methodology," *Computer-Aided Civil and Infrastructure Engineering*, 15(2), pp. 90–106.

Castillo, E., Gutierrez, J.M., Cobo, A. and Castillo, C. (2000b), "Some Learning Methods in Functional Networks," *Computer-Aided Civil and Infrastructure Engineering*, 15(6), pp. 426–438.

Cattan, J. and Mohammadi, J. (1997), "Analysis of Bridge Condition Rating Using Neural Networks," *Computer-Aided Civil and Infrastructure Engineering*, 12(6), pp. 419–429.

Chao, L. and Skibniewski, M.J. (1994), "Estimating Construction Productivity: by Neural Network-based Approach," *Journal of Computing in Civil Engineering*, 8(2), pp. 234–251.

Chao, L. and Skibniewski, M.J. (1995), "Neural Network Method of Estimating Construction Technology Acceptability," *Journal of Construction Engineering and Management*, 121(1), pp. 130–142.

Chen, H.M., Qi, G.Z., Yang, J.C.S. and Amini, F. (1995b), "Neural Network for Structural Dynamic Model Identification," *Journal of Engineering Mechanics*, 121(12), pp. 1377–1381.

Chen, H.M., Tsai, K.H., Qi, G.Z., Yang, J.C.S. and Amini, F. (1995a), "Neural Network for Structure Control," *Journal of Computing in Civil Engineering*, ASCE, 9(2), pp. 168–176.

Cheu, R.L. and Ritchie, S.G. (1995), "Automated Detection of Lane-Blocking Freeway Incidents Using Artificial Neural Networks," *Transportation Research – C*, 3(6), pp. 371–388.

Chikata, Y., Yasuda, N., Matsushima, M. and Kobori, T. (1998), "Inverse Analysis of Esthetic Evaluation of Planted Concrete Structures by Neural Networks," *Computer-Aided Civil and Infrastructure Engineering*, 13(4), pp. 255–264.

Chuang, P.H., Goh, A.T.C. and Wu, X. (1998), "Modeling the Capacity of Pin-Ended Slender Reinforced Concrete Columns Using Neural Networks," *Journal of Structural Engineering*, ASCE, 124(7), pp. 830–838.

Consolazio, G.R. (2000), "Iterative Equation Solver for Bridge Analysis Using Neural Networks," *Computer-Aided Civil and Infrastructure Engineering*, 15(2), pp. 107–119.

Coulibaly, P., Anctil, F. and Bobee, B. (2000), "Neural Network-Based Long Term Hydropower Forecasting System," *Computer-Aided Civil and Infrastructure Engineering*, 15(5), pp. 355–364.

Crespo, J.L. and Mora, E. (1995), "Neural Networks as Inference and Learning Engines," *Microcomputers in Civil Engineering*, 10(2), pp. 89–96.

Daoheng, S., Qiao, H. and Hao, X. (2000), "A Neurocomputing Model for the Elastoplasticity," *Computer Methods in Applied Mechanics and Engineering*, 182, pp. 177–186.

De Leon, G.P. and Knoke, J.R. (1995), "Probabilistic Analysis of Claims for Extensions in the Contract Time," *Computing in Civil Engineering*, ASCE, New York, Vol. 2, pp. 1513–1520.

Demeyer, S., Meijler, T.D., Nierstrasz, O. and Steyaert, P. (1997), "Design Guidelines for 'Tailorable' Frameworks," *Communications of the ACM*, 40(10), pp. 60–64.

Deo, M.C. and Chaudhari, G. (1998), "Tide Prediction Using Neural Networks," *Computer-Aided Civil and Infrastructure Engineering*, 13(2), pp. 113–120.

Deo, M.C., Venkat Rao, V. and Sakar, A. (1997), "Neural Networks for Wave Height Interpolation," *Microcomputers in Civil Engineering*, 12(3), pp. 217–225.

Dia, H. and Rose, G. (1997), "Development and Evaluation of Neural Network Freeway Incident Detection Models Using Field Data," *Transportation Research – C*, 5(5), pp. 313–331.

Du, Y.G., Sreekrishnan, T.R., Tyagi, R.D. and Campbell, P.G.C., (1994), "Effect of pH on Metal Solubilization from Sewage Sludge: A Neural-Net-Based Approach," *Canadian Journal of Civil Engineering*, 21, pp. 728–735.

Duan, L. and Chen, W.F. (1989), "Effective Length Factor for Columns in Unbraced Frames," *Journal of Structural Engineering*, ASCE, 115(1), pp. 149–165.

Elazouni, A.M., Nosair, I.A., Mohieldin, Y.A. and Mohamed, A.G. (1997), "Estimating Resource Requirements at Conceptual Design Stage Using Neural Networks," *Journal of Computing in Civil Engineering*, ASCE, 11(4), pp. 217–223.

Eldin, N.N. and Senouci, A.B. (1995), "A Pavement Condition Rating Model Using Backpropagation Neural Networks," *Microcomputers in Civil Engineering*, 10(6), pp. 433–441.

Elkordy, M.F., Chang, K.C. and Lee, G.C. (1994), "A Structural Damage Neural Network Monitoring System," *Microcomputers in Civil Engineering*, 9(2), pp. 83–96.

Eskandarian, A. and Thiriez, S. (1998), "Collision Avoidance Using Cerebral Model Arithmetic Computer Neural Network," *Computer-Aided Civil and Infrastructure Engineering*, 13(5), pp. 303–314.

Faghri, A. and Hua, J. (1995), "Roadway Seasonal Classification Using Neural Networks," *Journal of Computing in Civil Engineering*, ASCE, 9(3), pp. 209–215.

Faravelli, L.T. and Yao, T. (1996), "Use of Adaptive Networks in Fuzzy Control of Civil Structures," *Microcomputers in Civil Engineering*, 11(1), pp. 67–76.

Feng, C.-W., Liu, L. and Burns, S.A. (1997), "Genetic Algorithms to Solve Construction Time-Cost Trade-Off Problems," *Journal of Computing in Civil Engineering*, 11(3), pp. 184–189.

Feng, M.Q. and Bahng, E.Y. (1999), "Damage Assessment of Jacketed RC Columns Using Vibration Tests," *Journal of Structural Engineering*, ASCE, 125(3), pp. 265–271.

Fichman, R.G. and Kemerer, C.F. (1997), Object Technology and Reuse: Lessons Learned from Early Adopters," *Computer*, IEEE, 10, pp. 47–59.

Fischer, M. and Froese, T. (1996), "Examples and Characteristics of Shared Project Models," *Journal of Computing in Civil Engineering*, 10(3), pp. 174–182.

Fischer, M.A. and Aalami, F. (1996), "Scheduling with Computer-Interpretable Construction Method Models," *Journal of Construction Engineering and Management*, ASCE, 122(4), pp. 337–347.

Fowler, M. (1997), *Analysis Patterns: Reusable Objects Models*, Addison-Wesley Longman, Inc., Reading, Massachusetts.

Fowler, M. and Scott, K. (1997), *UML Distilled: Applying the Standard Object Modeling Language*, Addison-Wesley Longman, Inc., Reading, Massachusetts.

Froese, T. (1996), "Models of Construction Process Information," *Journal of Computing in Civil Engineering*, 10(3), pp. 183–193.

Froese, T. and Paulson, B.C. (1994), "OPIS: An Object Model-Based Project Information System," *Microcomputers in Civil Engineering*, 9, pp. 13–28.

Fukunaga, K. (1990), *Introduction to Statistical Pattern Recognition*, 2nd ed., Academic Press, Boston, Massachusetts.

Furukawa, T. and Yagawa, G. (1998), "Implicit Constitutive Modeling for Viscoplasticity Using Neural Networks," *International Journal for Numerical Methods in Engineering*, 43, pp. 195–219.

Furuta, H., He, J. and Watanabe, E. (1996), "A Fuzzy Expert System for Damage Assessment using Genetic Algorithms and Neural

Networks," *Microcomputers in Civil Engineering*, 11 (1), pp. 37–45.

Gagarin, N., Flood, I. and Albrecht, P. (1994), "Computing Truck Attributes with Artificial Neural Networks," *Journal of Computing in Civil Engineering*, ASCE, 8(2), pp. 179–200.

Gamma, E., Helm, R., Johnson, R. and Vlissides, J. (1995), *Design Patterns: Elements of Reusable Object-Oriented Software*, Addison-Wesley Publishing Company, Reading, Massachusetts.

Gangopadhyay, S., Gautam, T.R. and Gupta, A.D. (1999), "Subsurface Characterization Using Artificial Neural Network and GIS," *Journal of Computing in Civil Engineering*, ASCE, 13(3), pp. 153–161.

Ghaboussi, J., Pecknold, D.A., Zhang, M. and Haj-Ali, R.M. (1998), "Autoprogressive Training of Neural Network Constitutive Models," *International Journal for Numerical Methods in Engineering*, 42, pp. 105–126.

Ghaboussi, J. and Joghataie, A. (1995), "Active Control of Structure Using Neural Networks," *Journal of Engineering Mechanics*, 121(4), pp. 555–567.

Ghaboussi, J., Garrett, J.H. and Wu, X. (1991), "Knowledge-Based Modeling of Material Behavior with Neural Networks," *Journal of Engineering Mechanics*, ASCE, 117(1), pp. 132–153.

Goh, A.T.C. (1995), "Neural Networks for Evaluating CPT Calibration Chamber Test Data," *Microcomputers in Civil Engineering*, 10(2), pp. 147–151.

Golden, R.M. (1996), *Mathematical Methods for Neural Network Analysis and Design*, MIT Press, Cambridge, Massachusetts.

Grubert, J.P. (1995), "Prediction of Estuarine Instabilities with Artificial Neural Networks," *Journal of Computing in Civil Engineering*, ASCE, 9(4), pp. 266–274.

Gunaratnam, D.J. and Gero, J.S. (1994), "Effect of Representation on the Performance of Neural Networks in Structural Engineering Applications," *Microcomputers in Civil Engineering*, 9(2), pp. 97–108.

Guo, J.C.Y. (2001), "A Semi-Virtual Watershed Model by Neural Networks," *Computer-Aided Civil and Infrastructure Engineering*, 16(2).

Hajela, P. and Berke, L. (1991), "Neurobiological Computational Models in Structural Analysis and Design," *Computers and Structures*, 41(4), pp. 657–667.

Hajela, P. and Lee, E. (1997), "Topological Optimization of Rotorcraft Subfloor Structures for Crashworthiness Considerations," *Computers and Structures*, 64(1–4), pp. 65–76.

Handa, M. and Barcia, R.M. (1986), "Linear Scheduling Using Optimal Control Theory," *Journal of Construction Engineering and Management*, 112(3), pp. 387–393.

Hanna, A.S. and Senouci, A.B. (1995), "NEUROSLAB – Neural Network System for Horizontal Formwork Selection," *Canadian Journal of Civil Engineering*, 22, pp. 785–792.

Haykin, S. (1999), *Neural Networks: A Comprehensive Foundation*, 2nd ed., Prentice-Hall, Englewood Cliffs, New Jersey.

Hecht-Nielsen, R. (1987a), "Counterpropagation Networks," *Proceedings of the IEEE 1st International Conference on Neural Networks*, II, IEEE Press, New York, pp. 19–32.

Hecht-Nielsen, R. (1987b), "Counterpropagation Networks," *Applied Optics*, 26(23), pp. 4979–4985.

Hecht-Nielsen, R. (1988), "Application of Counterpropagation Networks," *Neural Networks*, 1(2), pp. 131–139.

Hegazy, T. and Moselhi, O. (1994), "Analogy-based Solution to Markup Estimation Problem," *Journal of Computing in Civil Engineering*, ASCE, 8(1), pp. 72–87.

Hegazy, T., Fazio, P. and Moselhi, O. (1994), "Developing Practical Neural Network Applications Using Backpropagation," *Microcomputers in Civil Engineering*, 9(2), pp. 145–159.

Hegazy, T., Tully, S. and Marzouk, H. (1998), "A Neural Network Approach for Predicting the Structural Behavior of Concrete Slabs," *Canadian Journal of Civil Engineering*, 25, pp. 668–677.

Hendrickson, C. and Au, T. (1989), *Project Management for Construction: Fundamental Concepts for Owners, Engineers, Architects, and Builders*, Prentice-Hall, Englewood Cliffs, NJ.

Hoit, M., Stoker, D. and Consolazio, G. (1994), "Neural Networks for Equation Renumbering," *Computers and Structures*, 52(5), pp. 1011–1021.

Hopfield, J.J. (1982), "Neural Networks and Physical Systems with Emergent Collective Computational Abilities," *Proceedings of the National Academy of Sciences*, 79, pp. 2554–2558.

Hopfield, J.J. (1984), "Neurons with Graded Response Have Collective Computational Properties Like Those of Two-State Neurons," *Proceedings of the National Academy of Sciences*, 81, pp. 3088–3092.

Huang, C.-C. and Loh, C.H. (2001), "Nonlinear Identification of Dynamic Systems using Neural Networks," *Computer-Aided Civil and Infrastructure Engineering*, 16(1), 28–41.

Hung, S.L. and Adeli, H. (1991a), "A Model of Perceptron Learning with a Hidden Layer for Engineering Design," *Neurocomputing*, 3, pp. 3–14.

Hung, S.L. and Adeli, H. (1991b), "A Hybrid Learning Algorithm for Distributed Memory Multicomputers," *Heuristics – The Journal of Knowledge Engineering*, 4(4), pp. 58–68.

Hung, S.L. and Adeli, H. (1993), "Parallel Backpropagation Learning Algorithm on Cray YMP8/864 Supercomputer," *Neurocomputing*, 5, pp. 287–302

Hung, S.L. and Adeli, H. (1994a), "Object-Oriented Backpropagation and Its Application to Structural Design," *Neurocomputing*, 6(1), pp. 45–55.

Hung, S.L. and Adeli, H. (1994b), "A Parallel Genetic/Neural Network Algorithm for MIMD Shared Memory Machines," *IEEE Transactions on Neural Networks*, 5, No. 6, pp. 900–909.

Hung, S.L. and Jan, J.C. (1999a), "MS_CMAC Neural Network Learning Model in Structural Engineering," *Journal of Computing in Civil Engineering*, ASCE, 13(1), pp. 1–11.

Hung, S.L. and Jan, J.C. (1999b), "Machine Learning in Engineering Analysis and Design – An Integrated Fuzzy Neural Network Learning Model," *Computer-Aided Civil and Infrastructure Engineering*, 14(3), pp. 207–219.

Hung, S.L., Kao, C.Y. and Lee, J.C. (2000), "Active Pulse Structural Control Using Artificial Neural Networks," *Journal of Engineering Mechanics*, 126(8), pp. 839–849.

Hurson, A.R., Pakzad, S. and Jin, B. (1994), "Automated Knowledge Acquisition in a Neural Network-Based Decision Support System for Incomplete Database Systems," *Microcomputers in Civil Engineering*, 9(2), pp. 129–143.

Hutchinson, J.M., Lo, A. and Poggio, T. (1994), "A Nonparametric Approach to Pricing and Hedging Derivative Securities Via Learning Networks," *MIT AI Memo*, No. 1471, Cambridge, Massachusetts.

Ibbs, C.W. (1997), "Quantitative Impacts of Project Change: Size Issues," *Journal of Construction Engineering and Management*, ASCE, 23(3), pp. 308–311.

Ivan, J.N. and Sethi, V. (1998), "Data Fusion of Fixed Detector and Probe Vehicle Data for Incident Detection," *Computer-Aided Civil and Infrastructure Engineering*, 13(5), pp. 329–337.

Jacobson, I., Christerson, M., Jonsson, P. and Overgaard, G. (1992), *Object-Oriented Software Engineering: A Use Case Driven Approach*, Addison-Wesley, New York.

Jayawardena, A.W. and Fernando, D.A.K. (1998), "Use of Radial Basis Function Type Artificial Neural Networks for Runoff Simulation," *Computer-Aided Civil and Infrastructure Engineering*, 13(2), pp. 91–99.

Jenkins, W.M. (1999), "A Neural Network for Structural Re-analysis," *Computers and Structures*, 72, pp. 687–698.

Jingui, L., Yunliang, D., Bin, W. and Shide, X. (1996), "An Improved Strategy for GAs in Structural Optimization," *Computers and Structures*, 61(6), pp. 1185–1191.

Johnson, R.E. (1997), "Frameworks = (Components + Patterns)," *Communications of the ACM*, 40(10), pp. 39–42.

Johnston, D.W. (1981), "Linear Scheduling Method for Highway Construction," *Journal of the Construction Division*, ASCE, 107(CO2), pp. 247–261.

Juang, C.H. and Chen, C.J. (1999), "CPT-based Liquefaction Evaluation Using Artificial Neural Networks," *Computer-Aided Civil and Infrastructure Engineering*, 14(3), pp. 221–229.

Juang, C.H., Ni, S.H. and Lu, P.C. (1999), "Training Artificial Neural Networks with the Aid of Fuzzy Sets," *Computer-Aided Civil and Infrastructure Engineering*, 14(6), pp. 407–415.

Kamarthi, S.V., Sanvido, V.E. and Kumara, S.R.T. (1992), "Neuroform – Neural Network System for Vertical Formwork

Selection," *Journal of Computing in Civil Engineering*, ASCE, 6(2), pp. 178–179.

Kaneta, T., Furusaka, S., Nagaoka, H., Kimoto, K. and Okamoto, H. (1999), "Process Model of Design and Construction Activities of a Building," *Computer-Aided Civil and Infrastructure Engineering*, 14(1), pp. 45–54.

Kang, H.-T. and Yoon, C.J. (1994), "Neural Network Approaches to Aid Simple Truss Design Problems," *Microcomputers in Civil Engineering*, 9(3), pp. 211–218.

Kao, J.-J. and Liao, Y.Y. (1996), "IAC Network for Composition of Waste Incineration Facility," *Journal of Computing in Civil Engineering*, ASCE, 10(2), pp. 168–171.

Kao, W.-M. and Adeli, H. (1997), "Distributed Object-Oriented Blackboard Model for Integrated Design of Steel Structures," *Microcomputers in Civil Engineering*, 12(2), pp. 141–155.

Karim, A. and Adeli, H. (1999a), "OO Information Model for Construction Project Management," *Journal of Construction Engineering and Management*, ASCE, 125(5), pp. 361–367.

Karim, A. and Adeli, H. (1999b), "CONSCOM: An OO Construction Scheduling and Change Management System," *Journal of Construction Engineering and Management*, ASCE, 125(5), pp. 368–376.

Karim, A. and Adeli, H. (1999c), "A New Generation Software for Construction Scheduling and Management," *Engineering, Construction and Architectural Management*, 6(4), pp. 380–390.

Karim, A. and Adeli, H. (1999d), "Global Optimum Design of Cold-Formed Steel Hat-Shape Beams," *Thin-Walled Structures*, 35(4), pp. 275–288.

Karim, A. and Adeli, H. (1999e), "Global Optimum Design of Cold-Formed Steel Z-Shape Beams," *Practice Periodical on Structural Design and Construction*, ASCE, 4(1), pp. 17–20.

Karim, A. and Adeli, H. (2000), "Global Optimum Design of Cold-Formed Steel I-Shape Beams," *Practice Periodical on Structural Design and Construction*, ASCE, 5(2), pp. 78–81.

Kartam, N. (1996), "Neural Network-Spreadsheet Integration for Earthmoving Operations," *Microcomputers in Civil Engineering*, 11(4), pp. 283–288.

Karunanithi, N., Grenney, W.J., Whitley, D. and Bovee, K. (1994), "Neural Networks for River Flow Prediction," *Journal of Computing in Civil Engineering*, ASCE, 8(2), pp. 201–220.

Kasperkiewicz, J., Racz, J. and Dubrawski, A. (1995), "HPC Strength Prediction Using Artificial Neural Networks," *Journal of Computing in Civil Engineering*, ASCE, 9(4), pp. 279–284.

Kim, J.T., Jung, H.J. and Lee, I.W. (2000a), "Optimal Structural Control Using Neural Networks," *Journal of Engineering Mechanics*, 126(2), pp. 201–205.

Kim, S.H., Yoon, C. and Kim, B.J. (2000b), "Structural Monitoring System Based on Sensitivity Analysis and a Neural Network," *Computer-Aided Civil and Infrastructure Engineering*, 15(4), pp. 309–318.

Kirkpatrick, S., Gelatt, C. and Vecchi, M. (1983), "Optimization by Simulated Annealing," *Science*, 220, 671–680.

Kishi, N., Chen, W.F. and Goto, Y. (1997), "Effective Length Factor of Columns in Semirigid and Unbraced Frames," *Journal of Structural Engineering*, ASCE, 123(3), pp. 313–320.

Kohonen, T. (1988), *Self-Organization and Associative Memory*, Springer-Verlag, Berlin.

Kolk, W.R. and Lerman, R.A. (1992), *Nonlinear System Dynamics*, Van Nostrand Reinhold, New York.

Kushida, M., Miyamoto, A. and Kinoshita, K. (1997), "Development of Concrete Bridge Rating Prototype Expert System with Machine Learning," *Journal of Computing in Civil Engineering*, ASCE, 11(4), pp. 238–247.

Leymeister, D.J., Shah, D. and Jain, S.K. (1993), "Computer Application in Analyzing Change Order Work," *Proceedings of the Fifth International Conference on Computing in Civil and Building Engineering*, Anaheim, CA, 7–9 June, ASCE, New York, Vol. 1, pp. 137–144.

Li, H., Shen, L.Y. and Love, P.D.E. (1999), "ANN-Based Mark-Up Estimation System with Self-Explanatory Capacities," *Journal of Construction Engineering and Management*, 125(3), pp. 185–189.

Li, S. (2000), "Global Flexibility Simulation and Element Stiffness Simulation in Finite Element Analysis with Neural Network,"

Computer Methods in Applied Mechanics and Engineering, 186, pp. 101–108.

Li, Z., Mu, B. and Peng, J. (2000), "Alkali-Silica Reaction of Concrete with Admixtures," *Journal of Engineering Mechanics,* 126(3), pp. 243–249

Liew, K.M. and Wang, Q. (1998), "Application of Wavelet Theory for Crack Identification in Structures," *Journal of Engineering Mechanics,* ASCE, 124(2), pp. 152–157.

Lingras, P. and Adamo, M. (1996), "Average and Peak Traffic Volumes: Neural Nets, Regression and Factor Approaches," *Journal of Computing in Civil Engineering,* ASCE, 10(4), pp. 300–306.

Liong, S.-Y., Lim, W.-H. and Paudyal, G.N. (2000), "River Stage Forecasting in Bangladesh: Neural Network Approach," *Journal of Computing in Civil Engineering,* ASCE, 14(1), pp. 1–8.

Liu, L., Burns, S.A. and Feng, C,-W. (1995), "Construction Time-Cost Trade-Off Analysis Using LP/IP Hybrid Method," *Journal of Construction Engineering and Management,* 121(4), pp. 446–454.

Liu, W. and James, C.S. (2000), Estimation of Discharge Capacity in Meandering Compound Channels Using Artificial Neural Networks," *Canadian Journal of Civil Engineering,* 27, pp. 297–308.

Maier, H.R. and Dandy, G.C. (1997), "Determining Inputs for Neural Networks Models of Multivariate Time Series," *Computer-Aided Civil and Infrastructure Engineering,* 12(5), pp. 353–368.

Manevitz, L., Yousef, M. and Givoli, D. (1997), "Finite Element Mesh Generation Using Self-Organizing Neural Networks," *Computer-Aided Civil and Infrastructure Engineering,* 12(4), pp. 233–250.

Marwala, T. (2000), "Damage Identification Using Committee of Neural Networks," *Journal of Engineering Mechanics,* 126(1), pp. 43–50.

Masri, S.F., Chassiakos, A.G. and Caughey, T.K. (1993), "Identification of Nonlinear Dynamic Systems Using Neural Networks," *Journal of Applied Mechanicss,* ASME, 60, pp. 123–133.

Masri, S.F., Nakamura, M, Chassiakos, A.G. and Caughey, T.K. (1996), "Neural Network Approach to Detection of Changes in

Structural Parameters," *Journal of Engineering Mechanics,* 122(4), pp. 350–360.

Masri, S.F., Smyth, A.W., Chassiakos, A.G., Caughey, T.K. and Hunter, N.F. (2000), "Application of Neural Networks for Detection of Changes in Nonlinear Systems," *Journal of Engineering Mechanics,* 126(7), pp. 666–676.

Masri, S.F., Smyth, A.W., Chassiakos, A.G., Nakamura, M. and Caughey, T.K. (1999), "Training Neural Networks by Adaptive Random Search Techniques," *Journal of Engineering Mechanics,* 125(2), pp. 123–132.

Mathew, A., Kumar, B., Sinha, B.P. and Pedreschi, R.F. (1999), "Analysis of Masonry Panel Under Biaxial Bending Using ANNs and CBR," *Journal of Computing in Civil Engineering,* ASCE, 13(3), pp. 170–177.

MathWorks, Inc. (1992), *MATLAB – High-Performance Numeric Computation and Visualization Software: User's Guide for UNIX Workstations,* MathWorks, Inc., Natick, Massachusetts.

Mattila, K. G. and Abraham, D.M. (1998), "Linear Scheduling: Past Research Efforts and Future Directions," *Engineering Construction and Architectural Management,* 5(3), pp. 294–303.

Mehrotra, K., Mohan, C.K. and Ranka, S. (1997), *Elements of Artificial Neural Networks,* MIT Press, Cambridge, Massachusetts.

Messner, J.I., Sanvido, V.E., Kumara, S.R.T. (1994), "StructNet: A Neural Network for Structural System Selection," *Microcomputers in Civil Engineering,* 9(2), pp. 109–118.

Microsoft (1997), *Microsoft Visual C++ MFC Library Reference,* Parts 1 and 2, Microsoft Press, Redmond, WA.

Mikami, I., Tanaka, S. and Hiwatashi, T. (1998), "Neural Network System for Reasoning Residual Axial Forces of High-Strength Bolts in Steel Bridges," *Computer-Aided Civil and Infrastructure Engineering,* 13(4), pp. 237–246.

Moder, J.J. and Phillips, C.R. (1970), *Project Management with CPM and PERT,* Van Nostrand Reinhold, New York.

Mohammed, H.A., Abd El Halim, A.O. and Razaqpur, A.G. (1995), "Use of Neural Networks in Bridge Management Systems," *Transportation Research Record,* No. 1490, pp. 1–8.

Moody, J. and Darken, C.J. (1989), "Fast Learning in Networks of Locally-Tuned Processing Units," *Neural Computation*, 1, pp. 281–294.

Moselhi, O, Hegazy, T. and Fazio, P. (1993), "DBID: Analogy-Based DSS for Bidding in Construction," *Journal of Construction Engineering and Management*, 119(3), pp. 466–479.

Moselhi, O., Hegazy, T. and Fazio, P. (1991), "Neural Networks as Tools in Construction," *Journal of Construction Engineering and Management*, 117(4), pp. 606–623.

Mukherjee, A. and Deshpande, J.M. (1995a), "Application of Artificial Neural Networks in Structural Design Expert Systems," *Computers and Structures*, 54(3), pp. 367–375.

Mukherjee, A. and Deshpande, J.M. (1995b), "Modeling initial Design Process Using Artificial Neural Networks," *Journal of Computing in Civil Engineering*, ASCE, 9(3), pp. 194–200.

Mukherjee, A., Deshpande, J.M. and Anmala, J. (1996), "Prediction of Buckling Load of Columns Using Artificial Neural Networks," *Journal of Structural Engineering*, 122(11), pp. 1385–1387.

Murtaza, M.B. and Fisher, D.J. (1994), "Neuromodex – Neural Network System for Modular Construction Decision," *Journal of Computing in Civil Engineering*, ASCE, 8(2), pp. 221–223.

Naylor, H.F.W. (1995), *Construction Project Management: Planning and Scheduling*, Delmar Publishers, Albany, New York.

Ni, S.H., Lu, P.C. and Juang, C.H. (1996), "A Fuzzy Neural Network Approach to Evaluation of Slope Failure Potential," *Microcomputers in Civil Engineering*, 11(1), pp. 59–66.

Nikzad, K., Ghaboussi, J. and Paul, S.L. (1996), "Actuator Dynamics and Delay Compensation Using Neurocontrollers," *Journal of Engineering Mechanics*, 122(10), pp. 966–975

Odenthal, G. and Quibeldey-Cirkel, K. (1997), "Using Patterns for Design and Documentation," *Proceedings of the 11th European Conference on Object-Oriented Programming (ECOOP '97)*, Jyvaskyla, Finland, June 1997, pp. 511–529.

Owusu-Ababia, S. (1998), "Effect of Neural Network Topology on Flexible Pavement Cracking Prediction," *Computer-Aided Civil and Infrastructure Engineering*, 13(5), pp. 349–355.

Pain, C.C., De Oliveira, C. R. E. and Goddard, A. J. H. (1999), "A Neural Network Graph Partitioning Procedure for Grid-Based Domain Decomposition," *International Journal for Numerical Methods in Engineering*, 44, pp. 593–613.

Pandey, P.C. and Barai, S.V. (1995), "Multilayer Perceptron in Damage Detection of Bridge Structures," *Computers and Structures*, 54(4), pp. 597–608.

Papadrakakis, M., Lagaros, N.D. and Tsompanakis, Y. (1998), "Structural Optimization Using Evolution Strategies and Neural Networks," *Computer Methods in Applied Mechanics and Engineering*, 156, pp. 309–333.

Papadrakakis, M., Papadopoulos, V. and Lagaros, N.D. (1996), "Structural Reliability Analysis of Elastic-Plastic Structures Using Neural Networks and Monte Carlo Simulation," *Computer Methods in Applied Mechanics and Engineering*, 136, pp. 145–163.

Park, D. and Rilett, L.R. (1999), "Forecasting Freeway Link Travel Times with a Multilayer Feedforward Neural Network," *Computer-Aided Civil and Infrastructure Engineering*, 14(5), pp. 357–367.

Park, H.S. and Adeli, H. (1995), "A Neural Dynamics Model for Structural Optimization – Application to Plastic Design of Structures," *Computers and Structures*, 57(1), pp. 391–400.

Park, H.S. and Adeli, H. (1997a), "Data Parallel Neural Dynamics Model for Integrated Design of Steel Structures," *Microcomputers in Civil Engineering*, 12(5), pp. 311–326.

Park, H.S. and Adeli, H. (1997b), "Distributed Neural Dynamics Algorithms for Optimization of Large Steel Structures," *Journal of Structural Engineering*, ASCE, 123(7), pp. 880–888.

Parvin, A. and Serpen, G. (1999), "Recurrent Neural Networks for Structural Optimization," *Computer-Aided Civil and Infrastructure Engineering*, 14(6), pp. 445–451.

Pena-Mora, F. and Hussein, K. (1999), "Interaction Dynamics in Collaborative Design Discourse," *Computer-Aided Civil and Infrastructure Engineering*, 14(3), pp. 171–185.

Poggio, T. and Girosi, P. (1990), "Networks for Approximation and Learning," *Proceedings of the IEEE*, 78(9), pp. 1481–1497.

Pompe, P.P.M. and Feedlers, A.J. (1997), "Using Machine Learning and Statistics to Predict Corporate Bankruptcy," *Microcomputers in Civil Engineering*, 12(4), pp. 267–276.

Press, W.H., Flannery, B.P., Teukolsky, S.A. and Vetterling, W.T. (1988), *Numerical Recipes in C: The Art of Scientific Computing*, Cambridge University Press, New York.

Rajasekaran, S., Febin, M.F. and Ramasamy, J.V. (1996), "Artificial Fuzzy Neural Networks in Civil Engineering," *Computers and Structures*, 61(2), pp. 291–302.

Razaqpur, A.G., Abd El Halim, A.O. and Mohamed, H.A. (1996), "Bridge Management by Dynamic Programming and Neural Networks," *Canadian Journal of Civil Engineering*, 23, pp. 1064–1069.

Reda, R.M. (1990), "RPM: Repetitive Project Modeling," *Journal of Construction Engineering and Management*, 116(2), pp. 316–330.

Roberts, C.A. and Attoh-Okine, N.O. (1998), "A Comparative Analysis of Two Artificial Neural Networks Using Pavement Performance Prediction," *Computer-Aided Civil and Infrastructure Engineering*, 13(5), pp. 339–348.

Rodriguez, M.J. and Sérodes, J. (1996), "Neural Network-Based Modelling of the Adequate Chlorine Dosage for Drinking Water Disinfection," *Canadian Journal of Civil Engineering*, 23, pp. 621–631.

Rogers, G.F. (1997), *Framework-Based Software Development in C++*, Prentice-Hall, Inc., Upper Saddle River, New Jersey..

Rogers, J.L. (1994), "Simulating Structural Analysis with Neural Network," *Journal of Computing in Civil Engineering*, ASCE, 8(2), pp. 252–265.

Rosenblatt, F. (1962), *Principle of Neurodynamics*, Spartan Books, New York.

Rudomin, P., Arbib, M.A., Cervantes-Perez, F. and Romo, R. (1993), *Neuroscience: From Neural Networks to Artificial Intelligence*, Springer-Verlag, Berlin.

Rumelhart, D.E., Hinton, G.E. and Williams, R.J. (1986), "Learning Internal Representation by Error Propagation," in Rumelhart, D.E., *et al.*, Eds., *Parallel Distributed Processing*, MIT Press, Cambridge, MA, pp. 318–362.

Russell, A.D. and Caselton, W.F. (1988), "Extensions to Linear Scheduling Optimization," *Journal of Construction Engineering and Management*, 114(1), pp. 36–52.

Russell, A.D. and Wong, W.C.M. (1993), "New Generation of Planning Structures," *Journal of Construction Engineering and Management*, 119(2), pp. 196–214.

Saito, M. and Fan, J. (2000), "Artificial Neural Network Based Heuristic Optimal Traffic Signal Timing," *Computer-Aided Civil and Infrastructure Engineering*, 15(4), pp. 281–291.

Samant, A. and Adeli, H. (2000), "Feature Extraction for Traffic Incident Detection Using Wavelet Transform and Linear Discriminant Analysis," *Computer-Aided Civil and Infrastructure Engineering*, 15(4), pp. 241–250.

Sankarasubramanian, G. and Rajasekaran, S. (1996), "Constitutive Modeling of Concrete Using a New Failure Criterion," *Computers and Structures*, 58(5), pp. 1003–1014.

Saunders, H. (1996), "Survey of Change Order Markups," *Practice Periodical on Structural Design and Construction*, ASCE, 1(1), pp. 15–19.

Savin, D., Alkass, S. and Fazio, P. (1996), "Construction Resource Leveling Using Neural Networks," *Canadian Journal of Civil Engineering*, 23, pp. 917–925

Savin, D., Alkass, S. and Fazio, P. (1998), "Calculating Weight Matrix of Neural Network for Resource Leveling," *Journal of Computing in Civil Engineering*, 12(4), pp. 241–248.

Sayed, T. and Abdelwahab, W. (1998), "Comparison of Fuzzy and Neural Classifiers for Road Accident Analysis," *Journal of Computing in Civil Engineering*, ASCE, 12(1), pp. 42–46.

Sayed, T. and Razavi, A. (2000), "Comparison of Neural and Conventional Approaches to Mode Choice Analysis," *Journal of Computing in Civil Engineering*, ASCE, 14(1), pp. 23–30.

Schmid, H.A. (1996), "Creating Applications from Components: A Manufacturing Framework Design," *IEEE Software*, 13(11), pp. 67–75.

Schmid, H.A. (1997), "Systematic Framework Design By Generalization," *Communications of the ACM*, 40(10), pp. 48–51.

Schmidt, D.C. and Fayad, M.E. (1997), Lessons Learned Building Reusable OO Frameworks for Distributed Software," *Communications of the ACM*, 40(10), pp. 85–87.

Seaburg, P.A. and Salmon, C.G. (1971), "Minimum Weight Design of Light Gage Steel Members," *Journal of the Structural Division*, ASCE, 97(ST1), pp. 203–222.

Selinger, S. (1980), "Construction Planning for Linear Projects," *Journal of Construction Division*, ASCE, 106(CO2), pp. 195–205.

Senouci, A. and Adeli, H. (2001), "Resource Scheduling Using Neural Dynamics Model of Adeli and Park," *Journal of Construction Engineering and Management*, ASCE, 127(1).

Shepherd, G. and Wingo, S. (1996), *MFC Internals – Inside the Microsoft Foundation Class Architecture*, Addison-Wesley Developers Press, Reading, Massachusetts.

Sinha, S.K. and McKim, R.A. (2000), "Artificial Neural Network for Measuring Organization Effectiveness," *Journal of Computing in Civil Engineering*, ASCE, 14(1), pp. 9–14.

Sonmez, R. and Rowings, J.E. (1998), "Construction Labor Productivity Modeling with Neural Networks," *Journal of Construction Engineering and Management*, 124(6), pp. 498–504.

Stavroulakis, G.E. and Antes, H. (1998), "Neural Crack Identification in Steady State Elastodynamics," *Computer Methods in Applied Mechanics and Engineering*, 165, pp. 129–146.

Stepanov, A. and Lee, M. (1995), *The Standard Template Library*, Hewlett-Packard Laboratories.

Stephens, J.E. and Vanluchene, R.D. (1994), "Integrated Assessment of Seismic Damage in Structures," *Microcomputers in Civil Engineering*, 9(2), pp. 119–128.

Stumpf, A.L., Ganeshan, R., Chin, S. and Liu, L.Y. (1996), "Object-Oriented Model for Integrating Construction Product and Process Information," *Journal of Computing in Civil Engineering*, 10(3), pp. 204–212.

Szewczyk, Z.P. and Hajela, P. (1994), "Damage Detection in Structures Based on Feature-Sensitive Neural Networks," *Journal of Computing in Civil Engineering*, ASCE, 8(2), pp. 163–178.

Szewczyk, Z.P. and Noor, A.K. (1996), "A Hybrid Neurocomputing/Numerical Strategy for Nonlinear Structural Analysis," *Computers and Structures*, 58(4), pp. 661–677.

Szewczyk, Z.P. and Noor, A.K. (1997), "A Hybrid Numerical/Neurocomputing Strategy for Sensitivity Analysis of Nonlinear Structures," *Computers and Structures*, 65(6), pp. 869–880.

Tang, Y. (1996a), "Active Control of SDF Systems Using Artificial Neural Networks," *Computers and Structures*, 60(5), pp. 695–703.

Tang, Y. (1996b), "New Algorithm for Active Structural Control," *Journal of Structural Engineering*, ASCE, 122(9), pp. 1081–1088.

Tashakori, A. and Adeli, H. (2001), "Optimum Design of Cold-Formed Steel Space Structures Using Neural Dynamics Model," *Journal of Structural Engineering*, ASCE, 127.

Tawfik, M., Ibrahim, A., Fahmy, H. (1997), "Hysteresis Sensitive Neural Network for Modeling Rating Curves," *Journal of Computing in Civil Engineering*, ASCE, 11(3), pp. 206–211

Teh, C.I., Wong, K.S., Goh, A.T.C. and Jaritngam, S. (1997), "Prediction of Pile Capacity Using Neural Networks," *Journal of Computing in Civil Engineering*, ASCE, 11(2), pp. 129–138.

Theocaris, P.S. and Panagiotopoulos, P.D. (1993), "Neural Networks for Computing in Fracture Mechanics. Methods and Prospects of Applications," *Computer Methods in Applied Mechanics and Engineering*, 106, pp. 213–228.

Theocaris, P.S. and Panagiotopoulos, P.D. (1995), "Generalised Hardening Plasticity Approximated via Anisotropic Elasticity: A Neural Network Approach," *Computer Methods in Applied Mechanics and Engineering*, 125, pp. 123–139.

Thirumalaiah, K. and Deo, M.C. (1998), "Real Time Flood Forecasting Using Neural Networks," *Computer-Aided Civil and Infrastructure Engineering*, 13(2), pp. 101–111.

Tikhonov, A.N. and Arsenin, V.Y. (1977), *Solution of Ill-Posed Problems*, W.H. Winston, Washington, DC.

Topping, B.H.V. and Bahreininejad, A. (1997), *Neural Computing for Structural Mechanics*, Saxe-Coburg Publications, Edinburgh, United Kingdom.

Topping, B.H.V., Khan, A.I. and Bahreininejad, A. (1997), "Parallel Training of Neural Networks for Finite Element Mesh Decomposition," *Computers and Structures*, 63(4), pp. 693–707.

Topping, B.H.V., Sziveri, J., Bahreininejad, A., Leite, J.P.B. and Cheng, B. (1998), "Parallel Processing, Neural Networks and Genetic Algorithms," *Advances in Engineering Software*, 29, pp. 763–786.

Turkkan, N. and Srivastava, N.K. (1995), "Prediction of Wind Load Distribution for Air-Supported Structures Using Neural Networks," *Canadian Journal of Civil Engineering*, 22, pp. 453–461.

UBC (1997), *Uniform Building Code, Vol. 2 – Structural Engineering Design Provisions,* International Conference of Building Officials, Whittier, California.

Vanluchene, R.D. and Sun, R. (1990), "Neural Networks in Structural Engineering," *Microcomputers in Civil Engineering*, 5(3), pp. 207–215.

Widrow, B. and Lehr, M.A. (1995), "Perceptrons, Adalines and Backpropagation," in Arbib, M.A., Ed., *The Handbook of Brain Theory and Neural Networks*, MIT Press, Cambridge, Massachusetts, pp. 719–724.

Williams, T.P. (1994), "Predicting Changes in Construction Cost Indexes Using Neural Networks," *Journal of Construction Engineering and Management*, ASCE, 120(2), pp. 306–320.

Williams, T.P. and Gucunski, N. (1995), "Neural Networks for Backcalculation of Moduli from SASW Tests," *Journal of Computing in Civil Engineering*, ASCE, 9(1), pp. 1–8.

Willis, E.M. (1986), *Scheduling Construction Projects*, John Wiley and Sons, Inc., New York.

Wu, X., Ghaboussi, J. and Garrett, J.H., Jr. (1992), "Use of Neural Networks in Prediction of Structural Damage," *Computers and Structures*, 42(4), pp. 649–659.

Yeh, I.-C. (1995), "Construction Site Layout Using Annealed Neural Networks," *Journal of Computing in Civil Engineering*, ASCE, 9(3), pp. 201–208.

Yeh, I.-C. (1998), "Structural Engineering Applications with Augmented Neuron Networks," *Computer-Aided Civil and Infrastructure Engineering*, 13(2), pp. 83–90.

Yeh, I.-C. (1999), "Design of High-Performance Concrete Mixture Using Neural Networks and Nonlinear Programming," *Journal of Computing in Civil Engineering*, ASCE, 13(1), pp. 36–42.

Yeh, Y.-C., Kuo, Y.-H. and Hsu, D.-S. (1993), "Building KBES for Diagnosing PC Pile with Artificial Neural Network," *Journal of Computing in Civil Engineering*, ASCE, 7(1), pp. 71–93.

Yen, G.G. (1994), "Identification and Control of Large Structures Using Neural Networks," *Computers and Structures*, 52(5), pp. 859–870.

Yen, G.G. (1996), "Distributive Vibration Control in Flexible Multibody Dynamics," *Computers and Structures*, 61(5), pp. 957–965.

Yu, G. and Adeli, H. (1991), "Computer-Aided Design Using Object-Oriented Programming Paradigm and Blackboard Architecture," *Microcomputer in Civil Engineering*, 6, pp. 177–189.

Yu, G. and Adeli, H. (1993), "Object-Oriented Finite Element Analysis Using an EER Model," *Journal of Structural Engineering*, ASCE, 119(9), pp. 2763–2781.

Yu, W.-W. (1991), *Cold-Formed Steel Design*, 2nd ed., John Wiley & Sons, Inc., New York.

Yun, C.B. and Bahng, E.Y. (2000), "Substructural Identification Using Neural Networks," *Computers and Structures*, 77, pp. 41–52.

Zadeh, L.A. (1970), "Decision Making in a Fuzzy Environment," *Management Sciences*, 17(4), pp. 141–164.

Zadeh, L. (1978), "Fuzzy Set as a Basis for a Theory of Possibility," *Fuzzy Sets and Systems*, 1(1), pp. 3–28.

SUBJECT INDEX

Milton Keynes UK
Ingram Content Group UK Ltd.
UKHW021626071024
449327UK00020BA/1207